An Isogeometric Approach to Beam Structures

Buntara S. Gan

An Isogeometric Approach to Beam Structures

Bridging the Classical to Modern Technique

Springer

Buntara S. Gan
College of Engineering, Department of Architecture
Nihon University
Koriyama, Fukushima, Japan

ISBN 978-3-319-56492-0 ISBN 978-3-319-56493-7 (eBook)
DOI 10.1007/978-3-319-56493-7

Library of Congress Control Number: 2017942943

Printed on acid-free paper

This Springer imprint is published by Springer Nature
The registered company is Springer International Publishing AG
The registered company address is: Gewerbestrasse 11, 6330 Cham, Switzerland

To

Rie, Kenta, Sūchan*, and* Shichan

Thank you for the love and encouragement over the years

Preface

This book is dedicated to bridging the classical beam theories and modern numerical simulation using the NURBS concept. If we observe carefully the state of the art and well-recognized finite element analysis (FEA) and computer-aided design (CAD), they are two different disciplines in engineering and science that do not go along with each other, until T.J.R. Hughes and co-workers in 2005 introduced the isogeometric analysis to bridge the gap between the CAD and the FEA approaches in which it gained significant thrusts to all research studies on this topic.

This book is intended for students in engineering and science who are beginning a series of courses in structures and mechanics, practitioners in companies who are starting to learn how to analyze the NURBS curves that they usually draw in their daily work, and researchers who are going to pursue or extend their research in a beam element using NURBS curves.

This book is written as a textbook or workbook with tutorials in how to get the problems solved. All the beam theories and NURBS concepts are accompanied by MATLAB® program lists which are intended for easy understanding and learning how the formulas work. Just like in computer science, we know the modern editor software now has been adopting the *WYSIWYG* system in which the displayed text can be edited in a form exactly resembling its appearance when printed on a piece of paper.

As for the organization of the book: In Chap. 1, the representation of curves on a plane is presented, starting from the parametric modeling of polynomial curves through several standard curves which are the foundation of the NURBS. All the methods described are accompanied by the MATLAB program and function lists. Some important concepts such as the control point, knot vector, weight, and rationalization are discussed. In Chap. 2, the numerical integration technique Gauss-Legendre quadrature is introduced. The Jacobian operator and curvature of a curve are explained by using examples with accompanying program lists. Chapter 3 reviews the classical theory of beams, how to get the governing equations, and how to formulate the beam in the finite element context. The applications of NURBS in the theory of beams and how to formulate the governing equation of various types of beams are described in Chap. 4. The core of this book is in Chap. 5, where a new proposed "Sandwich" condensation is introduced. The Sandwich condensation will condensate the increasing degree of freedom in the midspan of the beam due to adopting the NURBS, to a two-node six-degree-of-freedom standard beam element. Verification and validation are given. The program and function lists are provided. By condensing the NURBS-based beam element to a standard beam element, several merits will be gained. First, when we deal with nonlinear beam analyses where iterative calculations are needed, computing cost can be reduced significantly. Second, program and function lists can be *plugged in* seamlessly to the existing finite element codes either the in-house or commercial customizable software. In Chaps. 6, 7, and 8, the straight beam examples, circular curved beam examples, and free curved beam examples are given to demonstrate the advantage of the NURBS. Appendix A provides the element stiffness and mass matrices of the traditional straight beam element for verification purpose.

The writing of this book was my pleasant experience. I received benefits from my professional works, encouragements and supports from many colleagues as well as my students who have taught me how to explain to them complicated concepts in simple terms. While it is not possible to name all of them, without their helps and supports, it would not have been possible for me to finish the writing of this book.

My sincere thanks are due to my respected former Emeritus Professor Fumio Nishino in University of Tokyo, for his philosophies in teaching and research which have been useful in my professional life. Great appreciations are to Professor Mitsuharu Kurata in Nihon University, for constantly motivating me to find the "truth" in science. I wish to thank Dr. Nguyen Dinh Kien, my research colleague in VAST, for his fruitful ideas, encouragements, and support in my research works. I am indebted to Prof. Han Ay Lie in Universitas Diponegoro, for her continuous encouragements and inspiring ideas in doing my research works. I am also grateful to Prof. Lee Jaehong, Prof. Lee Seunghye, and Prof. Kim Nam Il in Sejong University for their hospitalities, collaborations, and many stimulating discussions concerning several research topics during my sabbatical leave. I would like to thank my former master's students, Mr. Shuichi Sekine and Mr. Yuto Konai, who have been working hard to modify and improve the program and work out some examples. I appreciate much the helps and supports from the Senior Publishing Editor Mr. Michael Luby, Project Coordinator Mr. Murugesan Tamilsevan and Project Manager Ms. Kirouchenadassou Sharmila during the preparation and publication of this book.

The readers are welcomed to use the MATLAB program and function lists freely. However, the author does not guarantee that the program and function lists are without any error. Also, the author is not responsible for any damage or loss by using the program or function lists. The reader is fully responsible for verifying before using the programs or function lists given in this book.

Most books are not free from errors, especially those with many mathematical equations and numbers. I wish to thank in advance those readers who are willing to draw attention to any typos and errors, using the e-mail address: [buntara@arch.ce.nihon-u.ac.jp].

Koriyama, Fukushima, Japan Buntara S. Gan

MATLAB® is a registered trademark of The MathWorks, Inc.

Contents

Chapter 1
Representation of Curves on a Plane

1.1 Parametric Modeling of Polynomial Curve

Parametric presentation of polynomial functions is one among the methods to represent curves on a plane in the geometric modeling. The parametric presentation of the polynomial is not very common in a real application; however, the understanding of parametric presentation is necessary to deal with the Isogeometric approach that we are going to study.

In the parametric form, each coordinate of a point on the curve is represented explicitly by a common parameter that controls the curvature. Therefore, by varying the value of the parameter, we can have a family of curves within the same order of the polynomial.

Suppose on the $x - y$ plane, the nth order of parametric polynomial takes the following form:

$$\begin{aligned} x &= a_0 + a_1\xi + a_2\xi^2 + \ . \ . \ . \ + a_n\xi^n \\ y &= b_0 + b_1\xi + b_2\xi^2 + \ . \ . \ . \ + b_n\xi^n \end{aligned} \tag{1.1}$$

Suppose we want to represent a curve as shown in Fig. 1.1 as a parametric polynomial form of (1.1). Firstly, we assume the curve is represented by the second order ($n = 2$) of polynomial function. Secondly, we need to define the common parameters ξ_1, ξ_2, ξ_3 to represent the parametric position of those $n + 1 = 3$ points. We can see here by defining different values of *positioning* those $n + 1$ points, we can have different curves of polynomial functions. Now, we need to obtain the values of a_0, a_1, a_2, b_0, b_1 and b_2 parameters to construct (1.1). To do this, we need to take the coordinates (x, y) of arbitrary three points (P_0, P_1, and P_2) resting on the curve. Usually, the first and second extremities (end points of the curve) are assigned to the 0th and nth points, respectively. By substitution, (1.1) results in

B.S. Gan, *An Isogeometric Approach to Beam Structures*,
DOI 10.1007/978-3-319-56493-7_1

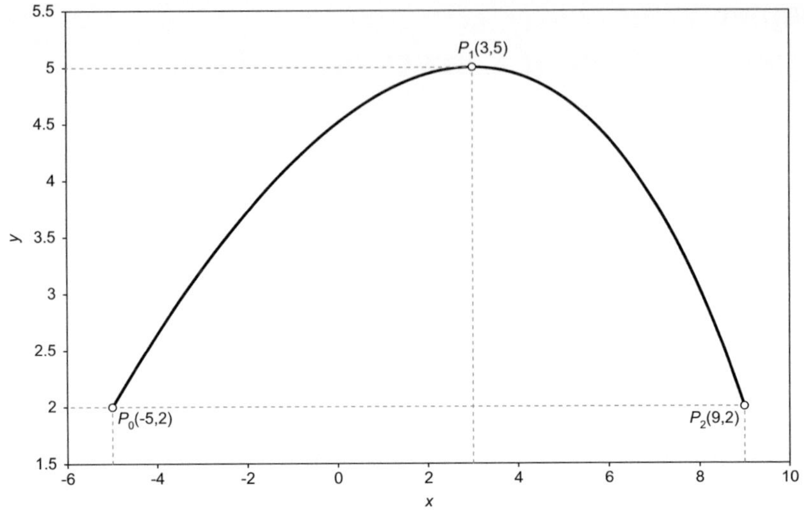

Fig. 1.1 Second-order parametric polynomial curve example

$$
\begin{aligned}
x_0 &= a_0 + a_1\xi_1 + a_2{\xi_1}^2 \\
x_1 &= a_0 + a_1\xi_2 + a_2{\xi_2}^2 \\
x_2 &= a_0 + a_1\xi_3 + a_2{\xi_3}^2 \\
y_0 &= b_0 + b_1\xi_1 + b_2{\xi_1}^2 \\
y_1 &= b_0 + b_1\xi_2 + b_2{\xi_2}^2 \\
y_2 &= b_0 + b_1\xi_3 + b_2{\xi_3}^2
\end{aligned}
\tag{1.2}
$$

In the matrix form

$$
\mathbf{x} = \mathbf{T}\ \boldsymbol{\alpha} \tag{1.3}
$$

where

$$
\begin{aligned}
\mathbf{x} &= [x_0\ x_1\ x_2\ y_0\ y_1\ y_2]^T \\
\boldsymbol{\alpha} &= [a_0 \quad a_1 \quad a_2 \quad b_0 \quad b_1 \quad b_2]^T \\
\mathbf{T} &= \begin{bmatrix} 1 & \xi_1 & \xi_1^2 & 0 & 0 & 0 \\ 1 & \xi_2 & \xi_2^2 & 0 & 0 & 0 \\ 1 & \xi_3 & \xi_3^2 & 0 & 0 & 0 \\ 0 & 0 & 0 & 1 & \xi_1 & \xi_1^2 \\ 0 & 0 & 0 & 1 & \xi_2 & \xi_2^2 \\ 0 & 0 & 0 & 1 & \xi_3 & \xi_3^2 \end{bmatrix}
\end{aligned}
$$

The vector $\boldsymbol{\alpha}$ can be obtained by multiplying the inverse of the matrix $\mathbf{T}^{-1}$ with the vector $\mathbf{x}$. Now let us consider, for example, the coordinates of three points taken from the curve in Fig. 1.1 are given by $P_0(-5, 2)$, $P_1(3, 5)$, and $P_3(9, 2)$. The common parametric values are initially guessed as $\xi_1 = -1$, $\xi_2 = 0$, and $\xi_3 = +1$. By substitution of the common parametric values into the matrix $\mathbf{T}$, then solving for vector $\boldsymbol{\alpha}$, results in $\boldsymbol{\alpha} = [3 \quad 7 \quad -1 \quad 5 \quad 0 \quad -3]^T$. By using the value of $\boldsymbol{\alpha}$, we can draw a curve as shown in Fig. 1.1 by varying the parameter ξ from -1 to $+1$ with a small interval.

Note: in this book, the interval of a parameter ξ is decided within the range of $-1 \leq \xi \leq 1$ so that it is consistent with the range of Gauss quadrature integration scheme that will be discussed in the subsequent following chapters. The Gauss quadrature integration scheme will be used for integrating all the matrix and vector quantities when we deal with finite element formulations for a beam developed by the Isogeometric approach.

1.1.1 Polynomial Program List

MATLAB code Polynomial.m solves the values of vector $\boldsymbol{\alpha}$, constructs the parametric polynomial equation in (1.1), and draws the polynomial curve as shown in Fig. 1.2a. Now, if we continue drawing all the curves by varying the value of parameter ξ, the range of -1 to $+1$, with 0.1 intervals, then we can get a second-order polynomial family that passes through those three points given.

```
% Polynomial.m
% Parametric construction of a 2nd order polynomial curve
% Nodal Coordinates: xi and yi
  xi=[-5 3 9];
  yi=[ 2 5 2];
  xyi=transpose([xi yi]);
% Parameter t
  t=[-1 0 1];
% [tm] matrix
  tm=[1 t(1) t(1)^2 0 0 0;        % x1
      1 t(2) t(2)^2 0 0 0;        % x2
      1 t(3) t(3)^2 0 0 0;        % x3
      0 0 0 1 t(1) t(1)^2;        % y1
      0 0 0 1 t(2) t(2)^2;        % y2
      0 0 0 1 t(3) t(3)^2];       % y3
% Coefficients a's and b's calculation
  ab=tm\xyi;
% Drawing graph
  ti=-1:0.02:1; n=size(ti,2);
  x=zeros(n);
  y=zeros(n);
```

```
for i=1:n
   x(i)=ab(1)+ab(2)*ti(i)+ab(3)*ti(i)^2;
   y(i)=ab(4)+ab(5)*ti(i)+ab(6)*ti(i)^2;
end
plot(x,y,'-k','LineWidth',2);
axis([-6 10 1.5 5.5]);
hold;
plot(xi,yi,'o','LineWidth',1,...
              'MarkerEdgeColor','k',...
              'MarkerFaceColor','w',...
              'MarkerSize',8);
xlabel('\itx'); ylabel('\ity');
text(xi(1),yi(1)-0.2,'\itP_{0}');
text(xi(2)-0.2,yi(2)+0.2,'\itP_{1}');
text(xi(3),yi(3)-0.2,'\itP_{2}');
text(2.2,4.7,'\it{\xi}_{2}=0');
```

1.1.2 Understanding Parametric Modeling

The most common geometric modeling that we learn in elementary school is how to draw the ordinate y from a given function corresponding to the abscissa x value. With this equation, we will only have one equation, and it is uniquely determined.

The concept of parametric modeling of a curve is rather difficult to imagine because of its implicit formulation relationship between x and y. However, the parametric modeling is quite easy to understand if we think both the abscissa x and the ordinate y in (1.1) are formulated by a common parameter ξ. The values of x and y are connected with one common parameter, ξ, at a time. Hence, we can have a relationship between the x and y, after we assign a value to parameter ξ. It is difficult to obtain a value of parameter ξ that can give exactly a pair of x and y if we do not have the coefficients in (1.2). We will face difficulties if we have to transform the (1.1) into the "y as a function of x," especially for higher-order degree, where the elimination of parameter ξ from both equations becomes tedious works by hand, indeed. But in the geometric computation by using a computer, the parametric modeling suits well the representation of curves numerically.

To illustrate how the parameter ξ works, after we solved for vector $\boldsymbol{\alpha}$, we can draw both x and y graphs as functions of parameter ξ independently. If we draw both graphs to be on the right for ξ-y and at the bottom for ξ-x of the constructed x-y curve, then we can have all the graphs connected as shown in Fig. 1.3. In fact, for each value of parameter ξ, there is a pair of two corresponding values of x and y. Thus, implicitly both values of x and y are connected indirectly through the parameter ξ. From there we can observe distinctly that both curves x and y are not in the same type of second order of polynomial.

The question now is what happened if we guessed the parametric vector not as $\Xi = \begin{bmatrix} -1 & 0 & 1 \end{bmatrix}^T$? Will we get the same answer? No. Actually, we made a wrong

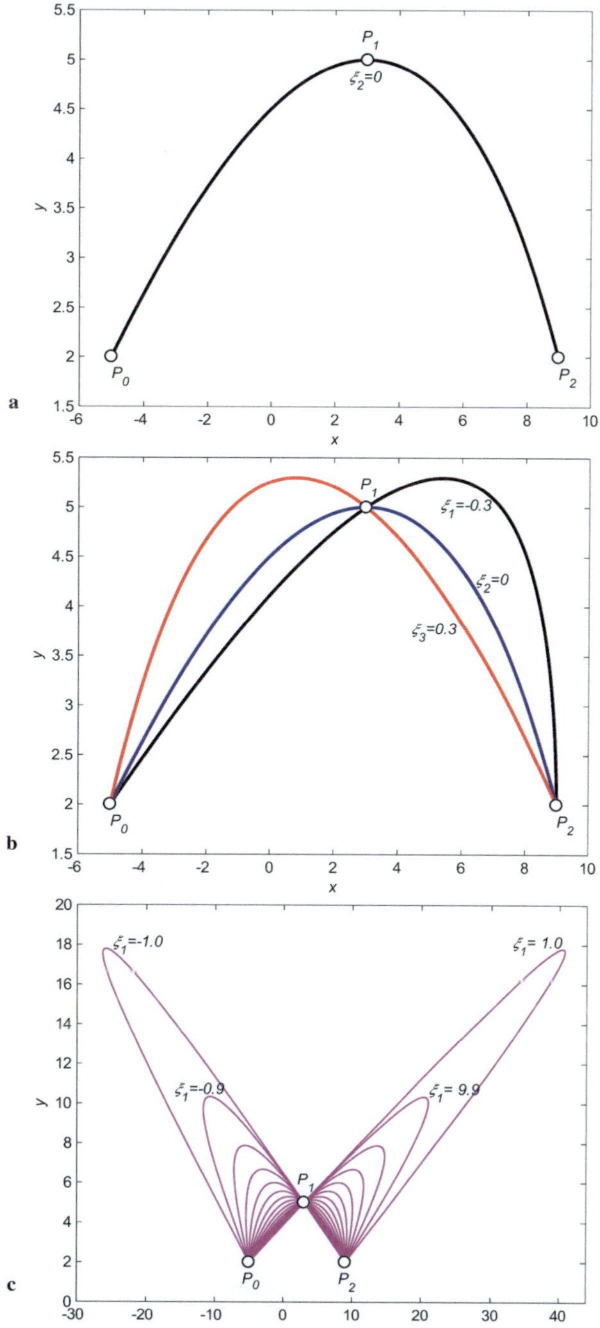

Fig. 1.2 (**a**) A second-order parametric polynomial curve constructed by given coordinates of three points (P_0, P_1, P_2) and defined parametric vector of $\boldsymbol{\Xi} = [-1 \quad 0 \quad 1]^T$. (**b**) The P_1 is assigned by different values of midpoint parametric value ξ_2. (**c**) The family of second-order parametric polynomial curve constructed by varying the midpoint parametric value in the range of $-1 \leq \xi_2 \leq 1$

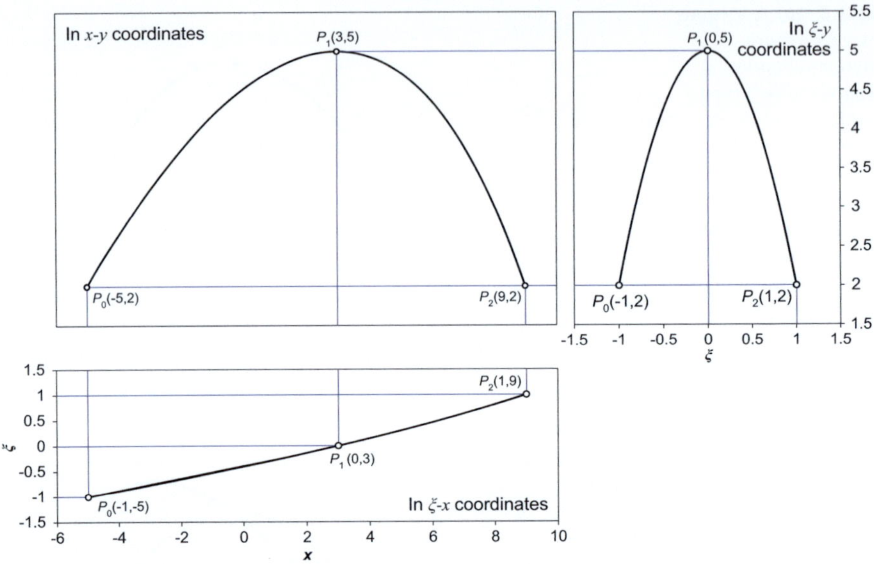

Fig. 1.3 A second-order parametric polynomial curve created by given coordinates of three points (P_0, P_1, P_2) and parametric vector $\Xi = [-1 \quad 0 \quad 1]^T$ which are drawn in (*x*, *y*), (*ξ*, *y*), and (*ξ*, *x*) graphs

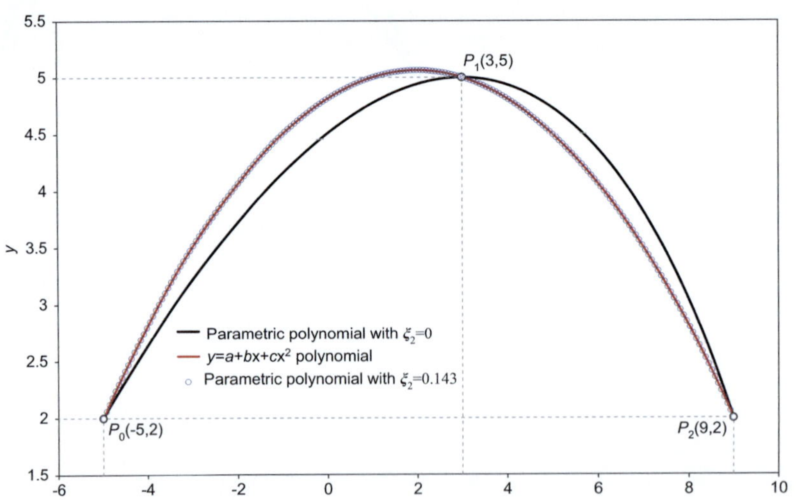

Fig. 1.4 Second-order parametric polynomial curve constructed by given coordinates of three points (P_0, P_1, P_2) and defined parametric vector of $\Xi = [-1 \quad 0 \quad 1]^T$ and $\Xi = [-1 \quad 0.143 \quad 1]^T$

guess! The two parametric values of $\xi_1 = -1$, $\xi_3 = 1$ which were guessed at both extremities of the curve are correct. But, the second value $\xi_2 = 0$ which is the midpoint of both values in the parametric modeling was actually not the *right* value. Actually, the parameter ξ must be interpolated along the length of the path of the curve; unless the curve is a linear one, the relationship between the parameter ξ is not linear either with the coordinate x or y.

Now, let us use the ordinary nonparametric polynomial for solving three simultaneous equations from the given three points, and we can get the following solution:

$$y = 4.813 + 0.250x - 0.063x^2$$

The solution is a unique one since it will give a single curve. If we draw this curve together with the one we computed by using the guessed parametric vector where the second value $\xi_2 = 0$, then we can see both curves as shown in Fig. 1.4 (red line from Fig. 1.1 and black line from Fig. 1.2) are not in coincidence. Why? Because we made a different (not "wrong") guess of $\xi_2 = 0$, which means the second point of the parameter was not in a correct position along the path of the curve. To get the similar curve with the solution, we have to set the value of $\xi_2 = 0.143$ for the second sampling point $P_1(3, 5)$. Then, which one is correct? The answer is that both are correct. Because we can have an infinite number of second order of polynomial curves if we want to draw the curves by using the parametric modeling concept. For higher order of the polynomial, which is difficult to determine, a unique single value of ξ, an additional algorithm that can calculate the length of the curve to select a correct sampling point, is the only choice to have the real curve that we wanted to construct.

Although the curve construction by using parametric polynomial modeling has some disadvantages to representing a curve in geometric modeling, they are still very useful in generating a variety form of polynomial curves because the polynomial forms are simple and mathematically well understood. The parametric form of a curve is suitable for designing and representing free geometric form by using a computer. In the remaining parts of this chapter, we will deal mostly with parametric modeling.

In this book, we are not going to go into much detail on the geometry modeling of curves. For more informative journals and standard textbooks of the topic, the reader can find more detail and other aspects in geometric modeling in Bajaj et al. (1995), Bernstein (1912), Faux and Pratt (1981), Mortenson (1985), Hoffmann (1989) and Beach (1991), Lei and Wang (2009), Piegl and Tiller (1997, 2003), and Lorentz (1986).

1.2 Bézier Curve

In the previous section, we found that the parametric polynomial is widely used a class of functions in geometric modeling. Although they are easy, efficient and accurately processed in a computer, there are some important curve types which cannot be precisely represented by using the parametric polynomials. Those unordinary types of curves must be represented by a system that consists of polynomials. In this section, we learn a method so-called the Bézier curve. Our discussion of the Bézier curve is only the essential part that related to the parametric modeling; in more detail the reader should consult other references (Bajaj et al. 1995; Bézier 1972,1986; Chang and Wu 1981; Chou 1995; Farin 1993; Gordon and Riesenfeld 1974; Piegl and Tiller 2003; Hoschek and Lasser 1993; Rogers and Adams 1990; Yamaguchi 1988).

The nth-order Bézier curve is given by

$$\begin{aligned} x &= \sum_{i=0}^{n} B_{i,n}(\xi)\; P_{xi} \\ y &= \sum_{i=0}^{n} B_{i,n}(\xi)\; P_{yi} \end{aligned} \qquad -1 \le \xi \le 1 \tag{1.4}$$

In (1.4), variables P_{xi}, P_{yi} are called *control points* in the x-y plane. The $B_{i,n}(\xi)$ are ξ-parametric based basis functions which are also known as the nth-order Bernstein polynomials (Bernstein 1912; Lorentz 1986). The Bernstein polynomials are defined by

$$B_{i,n}(\xi) = \frac{n!}{i!(n-i)!}\left(\frac{1+\xi}{2}\right)^{i} \times \left(\frac{1-\xi}{2}\right)^{n-i} \tag{1.5}$$

1.2.1 BernsteinBasis Program List

MATLAB code BernsteinBasis.m and function Bernstein.m (Sect. 1.2.2) draws the Bernstein curves for $n=2$ and $n=4$ as shown in Fig. 1.5 which are based on the (1.5). Notice that we can have more polynomial forms to compose a curve by following the Bernstein polynomial rules. Because we have $n+1$ curves as the basis functions corresponding to each $n+1$ control point to draw a polynomial curve with a predefined order of n.

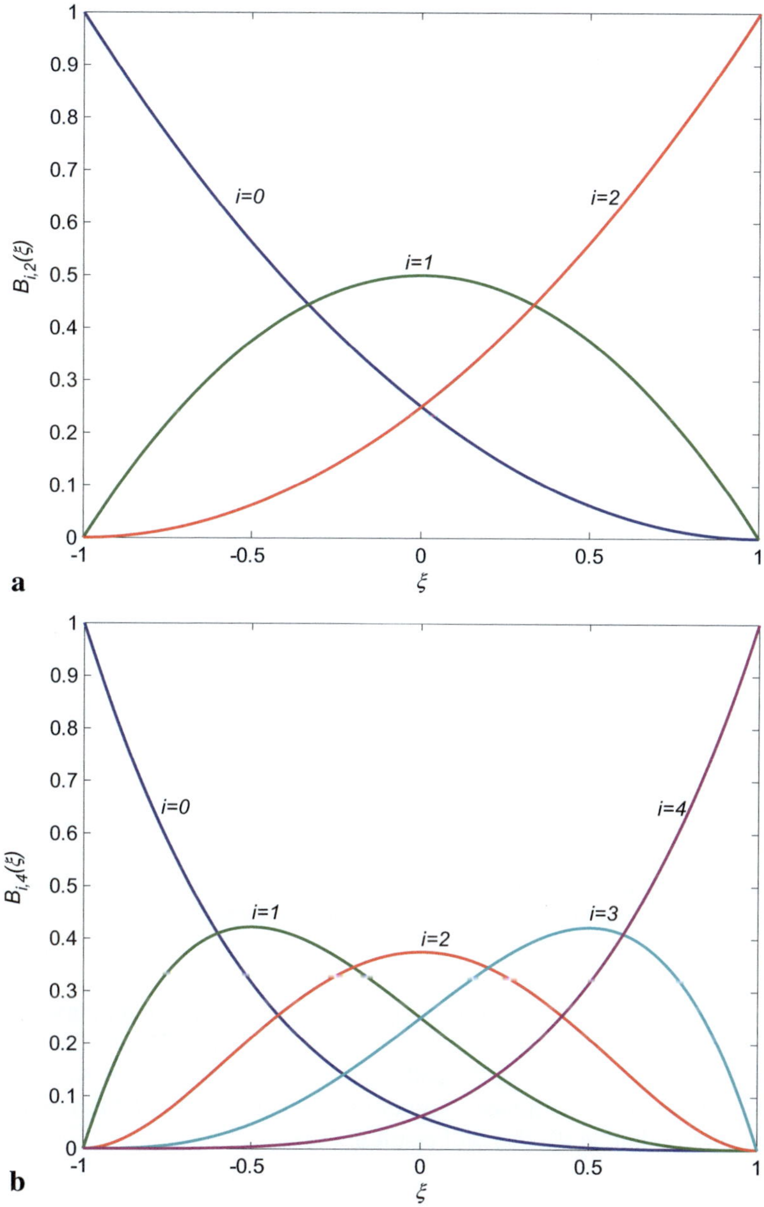

Fig. 1.5 (**a**) Second order of Bernstein basis functions ($n = 2$). (**b**) Fourth order of Bernstein basis functions ($n = 4$)

```
% BernsteinBasis.m
% Drawing the Bernstein basis functions
% n = order of curve
% t = -1 to +1, incremented by dt
n=2; % 2nd/4th order of polynomial
dt=1/50;
t=-1:dt:1;
y=zeros(size(t));
for i=0:n
  for dt=1:size(t,2)
    y(dt)=Bernstein(n,i,t(dt));
  end
  plot(t,y,'LineWidth',2);
  hold all;
end
xlabel('X'); ylabel('Y');
```

1.2.2 Bernstein Function List

```
function Brnbasis = Bernstein(n,i,t)
% Bernstein.m
% Bernstein basis functions
% n = order of Bezier curve
% t = -1 to +1, parameter
Brnbasis=factorial(n)/factorial(i)/...
         factorial(n-i)*(1/2+t/2)^i*(1/2-t/2)^(n-i);
return
```

Similar with the previous section, suppose we want to represent a curve as shown in Fig. 1.1 in a Bézier second-order polynomial form of (1.4). We need to obtain the control points $P_{x0}, P_{x1}, P_{x2}, P_{y0}, P_{y1}$ and P_{y2} to construct the equation in (1.4). To do this, we use the same coordinates (x,y) of three points (P_0, P_1 and P_2) in Sect. 1.1, and substituting into (1.4), that will result in (1.6).

$$
\begin{aligned}
x_0 &= B_{0,2}(\xi_1)\ P_{x0} + B_{1,2}(\xi_1)\ P_{x1} + B_{2,2}(\xi_1)\ P_{x2} \\
x_1 &= B_{0,2}(\xi_2)\ P_{x0} + B_{1,2}(\xi_2)\ P_{x1} + B_{2,2}(\xi_2)\ P_{x2} \\
x_2 &= B_{0,2}(\xi_3)\ P_{x0} + B_{1,2}(\xi_3)\ P_{x1} + B_{2,2}(\xi_3)\ P_{x2} \\
y_0 &= B_{0,2}(\xi_1)\ P_{y0} + B_{1,2}(\xi_1)\ P_{y1} + B_{2,2}(\xi_1)\ P_{y2} \\
y_1 &= B_{0,2}(\xi_2)\ P_{y0} + B_{1,2}(\xi_2)\ P_{y1} + B_{2,2}(\xi_2)\ P_{y2} \\
y_2 &= B_{0,2}(\xi_3)\ P_{y0} + B_{1,2}(\xi_3)\ P_{y1} + B_{2,2}(\xi_3)\ P_{y2}
\end{aligned}
\tag{1.6}
$$

Then, the control points can be computed from the sampling data of $n + 1$ points taken from the curve of concern. From those control points, we can draw the curve by using (1.4). In a similar fashion with the previous section, the difficulties for determining the value of parameter ξ for the sampling points for an existing curve remain the main issue in this Bézier curve construction. As long as we use the parametric modeling for drawing a curve, this issue will never disappear, because, in the parametric modeling, the solutions are not unique.

In the matrix form

$$
\mathbf{x} = \mathbf{T}\ \boldsymbol{\alpha} \tag{1.7}
$$

where

$$
\begin{aligned}
\mathbf{x} &= \begin{bmatrix} x_0 & x_1 & x_2 & y_0 & y_1 & y_2 \end{bmatrix}^T \\
\mathbf{P} &= \begin{bmatrix} P_{x0} & P_{x1} & P_{x2} & P_{y0} & P_{y1} & P_{y2} \end{bmatrix}^T \\
\mathbf{T} &= \begin{bmatrix}
B_{0,2}(\xi_1) & B_{1,2}(\xi_1) & B_{2,2}(\xi_1) & 0 & 0 & 0 \\
B_{0,2}(\xi_2) & B_{1,2}(\xi_2) & B_{2,2}(\xi_2) & 0 & 0 & 0 \\
B_{0,2}(\xi_3) & B_{1,2}(\xi_3) & B_{2,2}(\xi_3) & 0 & 0 & 0 \\
0 & 0 & 0 & B_{0,2}(\xi_1) & B_{1,2}(\xi_1) & B_{2,2}(\xi_1) \\
0 & 0 & 0 & B_{0,2}(\xi_2) & B_{1,2}(\xi_2) & B_{2,2}(\xi_2) \\
0 & 0 & 0 & B_{0,2}(\xi_3) & B_{1,2}(\xi_3) & B_{2,2}(\xi_3)
\end{bmatrix}
\end{aligned}
$$

MATLAB code BezierCurve.m uses function Bernstein.m (Sect. 1.2.2) and solves the values of the vector **P**, further constructs the Bézier polynomial equation in (1.4), and finally draws the Bézier curve as shown in Fig. 1.6.

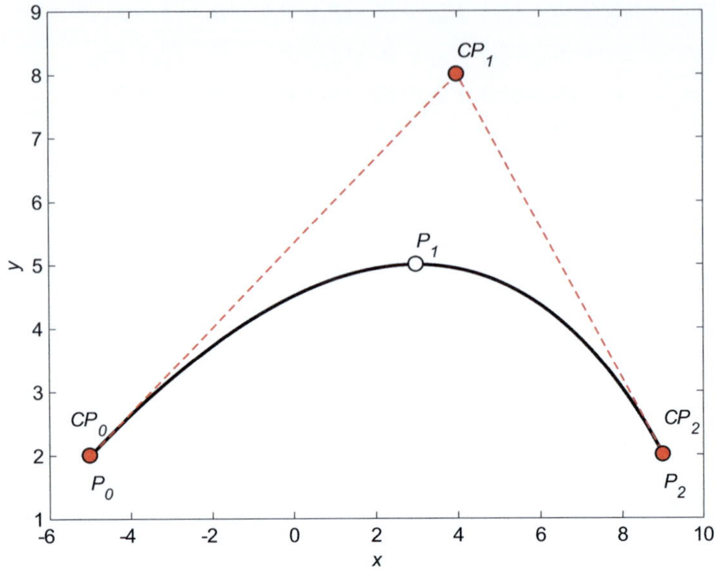

Fig. 1.6 A second-order Bézier polynomial curve constructed from a given coordinates of three points (P_0, P_1, P_2) and assumed parametric vector $\Xi = [-1 \quad 0 \quad 1]^T$

1.2.3 BezierCurve Program List

```
% BezierCurve.m
% Drawing the Bezier curve
% n = order of Bezier curve
% t = -1 to +1, incremented by dt
clear all;
n=2;
% Nodal coordinates: xi and yi
xi=[-5 3 9];
yi=[ 2 5 2];
xyi=transpose([xi yi]);
% Parameter t
t=[-1 0 1];
nt=size(t,2);
tmi=zeros(nt,n+1);
for i=0:n
  for dt=1:nt
    tmi(dt,i+1) = Bernstein(n,i,t(dt));
  end
end
```

```
% [tm] matrix
tm=[tmi zeros(3,3);zeros(3,3) tmi];
% Coefficients Px's and Py's calculation
pxy=tm\xyi;
px0=[pxy(1:nt)]; % n+1 control points
py0=[pxy(nt+1:2*nt)]; % n+1 control points
px=px0; py=py0;
dt=1/40; % steps
ti=-1:dt:1;
nt=size(ti');
curvx=zeros(nt);
curvy=zeros(nt);
for i=0:n
  for dt=1:nt
    Brnbasis = Bernstein(n,i,ti(dt));
    curvx(dt)=curvx(dt)+Brnbasis*px(i+1);
    curvy(dt)=curvy(dt)+Brnbasis*py(i+1);
  end
end
plot(curvx,curvy,'-k','LineWidth',2);
axis([-6 10 1 9]);
hold all;
plot(xi,yi,'ko','LineWidth',1,'MarkerEdgeColor',...
           'k','MarkerFaceColor','w','MarkerSize',8);
plot(px,py,'--ro','LineWidth',1,'MarkerEdgeColor',...
           'k','MarkerFaceColor','r','MarkerSize',8);
hold all;
xlabel('\itx'); ylabel('\ity');
text(px(1)-0.5,py(1)+0.5,'\itCP_{0}');
text(px(2),py(2)+0.5,'\itCP_{1}');
text(px(3),py(3)+0.5,'\itCP_{2}');
text(xi(1),yi(1)-0.5,'\itP_{0}');
text(xi(2),yi(2)+0.5,'\itP_{1}');
text(xi(3),yi(3)-0.5,'\itP_{2}');
```

The coordinates of the control points obtained from the calculation are $CP_0(-5,2)$, $CP_1(4,8)$, and $CP_2(9,2)$.

Note that the Bézier curves are suitable for intuitive design by using free-form curves. This is because the *control points* give the designer more handles to harness the curve creations. Now, let us explore the capability of Bézier polynomial in creating fascinating curves that the standard polynomial functions cannot do.

1.2.4 What is a Basis Function?

A basis function is a single curve element which is used as a basis of linear combination to construct a particular curve. We can see in Fig. 1.5a ($n = 2$) and Fig. 1.5b ($n = 4$) the number of basis function equals to the order polynomial of the curve plus one. Equation (1.5) defines the Bernstein polynomial that is used as the basis functions for constructing the Bézier curve. The series of basis functions have standard polynomial forms which are symmetric at the midpoint of the parameter ξ interval.

These basis functions will give a value less than one. But, if we sum all the basis functions along the parameter ξ, we can get a constant value of one. Basis functions having this property are said to be having a partition of unity. The partition of unity is a property which is required for selecting a shape function to be adopted in the finite element formulations.

The basis function is associated with a control point in calculating the linear combination of (1.4).

1.2.5 The Roles of a Control Point

A control point plays a major role in constructing a Bézier curve. If we observe the relationship between a basis function and a control point in (1.4), the ith control point is a scalar multiplier of the ith basis function. We call the control point a "*controller*" to a corresponding basis function. Then, we need $n + 1$ number of control points for the $n + 1$ basis functions to construct a Bézier curve.

However, we can see in Fig. 1.7 only the control points CP_0 and CP_2 are on the curve drawn by the formula, but not for CP_1. This is because the first ($i = 0$) and the last ($i = n$) basis functions give the values of $B_{\mathrm{i}} = 1$ at $\xi_0 = -1$ and $\xi_n = +1$, respectively. Hence, the computed x_0, x_2, y_0, and y_2 in (1.4) are indeed the coordinates of the CP_0 and CP_2 themselves, respectively. There is a small chance that we can get the second control point CP_1 laying on the curve because we can see from Fig. 1.7 that the coordinates of the second control point are a summation of the three nonzero basis functions at $\xi_0 = -1$ multiplied by their corresponding three control points. In fact, if we see carefully Fig. 1.7, the Bézier curve is constructed simultaneously by the three control points, not only the second one. In other words, all control points contribute to Bézier curve construction according to their corresponding basis function provided. For higher nth order of Bézier polynomial, we have $n + 1$ sets of basis functions and their corresponding control points.

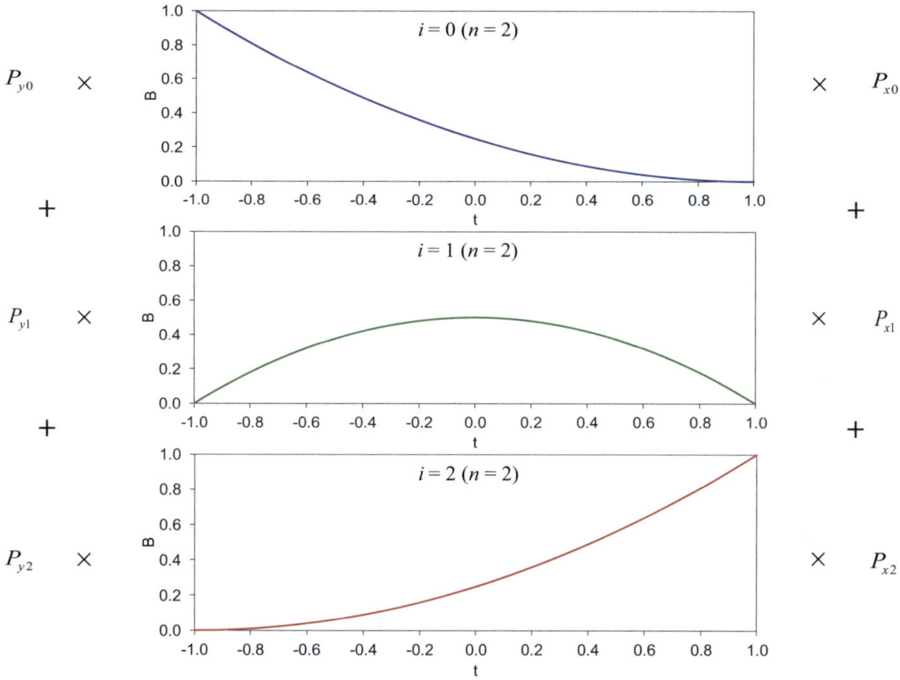

Fig. 1.7 Second-order Bernstein polynomial basis functions which are multiplied by their corresponding control points and summed up to obtain the x and y coordinates as depicted in Fig. 1.6

1.2.6 BezierCurveCP Program List

MATLAB code BezierCurveCP.m uses function Bernstein.m (Sect. 1.2.2) and draws the Bézier curves as shown in Fig. 1.8.

```
% BezierCurveCP.m
% Drawing the Bezier curves from given control points
% n = order of Bezier curve
% t = -1 to +1, incremented by dt
clear all;
n=3;
px0=[ 0 8 -6 2]; % n+1 control points
py0=[ 0 6  6 0]; % n+1 control points
px=px0; py=py0;
dt=1/50; % steps
t=-1:dt:1;
nt=size(t');
curvx=zeros(nt);
curvy=zeros(nt);
```

```
for i=0:n
  for dt=1:nt
    Brnbasis = Bernstein(n,i,t(dt));
    curvx(dt)=curvx(dt)+Brnbasis*px(i+1);
    curvy(dt)=curvy(dt)+Brnbasis*py(i+1);
  end
end
plot(curvx,curvy,'-k','LineWidth',2);
axis([min(px0)-2 max(px0)+2 min(py0)-2 max(py0)+2]);
hold all;
plot(px,py,'--ro','LineWidth',1,'MarkerEdgeColor',...
           'k','MarkerFaceColor','r','MarkerSize',8);
hold all;
xlabel('X'); ylabel('Y');
text(px(1)-0.5,py(1)-0.5,'CP0');
text(px(2)-0.5,py(2)+0.5,'CP1');
text(px(3)-0.5,py(3)+0.5,'CP2');
text(px(4)-0.5,py(4)-0.5,'CP3');
text(0.5,5,'Loop');
```

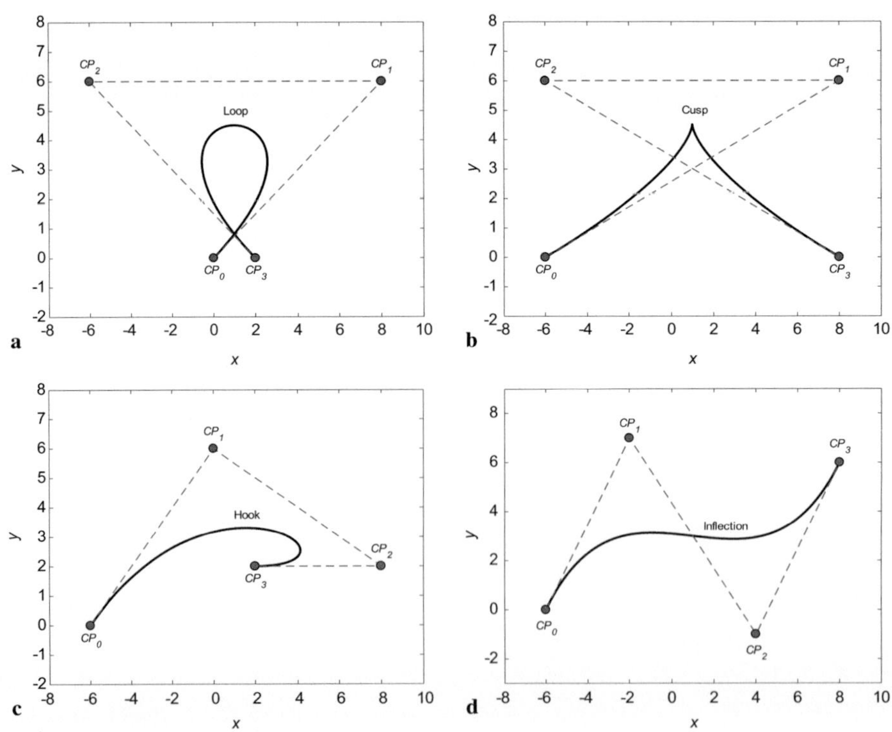

Fig. 1.8 Third-order Bézier polynomial curve examples by using four control points to create (**a**) Loop [(0,0); (8,6); (−6,6); (2,0)], (**b**) Cusp [(−6,0); (8,6); (−6,6); (8,0)], (**c**) Hook [(−6,0); (0,6); (8,2); (2,2)], (**d**) Inflection [(−6,0); (−2,7); (4,-1); (8,6)]

1.2.7 Rational Bézier Curve

There are many applications requiring the family of functions of curves which have been normalized. The Bézier curves can be represented by *rational functions* which are defined as the ratio between the family of the polynomials. We can construct such curves by dividing each polynomial by the summation form of functions as a denominator.

The *n*th order of rational Bézier curve is defined by

$$\begin{aligned} x &= \sum_{i=0}^{n} R_{i,n}(\xi)\ P_{xi} \\ y &= \sum_{i=0}^{n} R_{i,n}(\xi)\ P_{yi} \end{aligned} \qquad -1 \le \xi \le 1 \tag{1.8}$$

where

$$R_{i,n}(\xi) = \frac{B_{i,n}(\xi)\ w_i}{\sum_{j=0}^{n} B_{j,n}(\xi)\ w_j} \qquad w_i > 0$$

The w_i are *weights* of control points. The $R_{i,n}(\xi)$ are the rational basis functions for the Bézier curve.

Now, let us see if we use the same sampling point data from Sect. 1.1 with the weights of $w_0 = 1$, $w_1 = 2$, and $w_2 = 3$.

Similar to the previous sections, suppose we want to represent a curve as shown in Fig. 1.1 by a rational Bézier second-order polynomial form of (1.8). Now, we need to obtain the control points $P_{x0}, P_{x1}, P_{x2}, P_{y0}, P_{y1}$ and P_{y2} to construct the equation in (1.8). To do this, we use the same coordinates (*x*,*y*) of three points (P_0, P_1, and P_2) in Sect. 1.1 and substituting into (1.8) results in

$$\begin{aligned} x_0 &= R_{0,2}(\xi_1)\ P_{x0} + R_{1,2}(\xi_1)\ P_{x1} + R_{2,2}(\xi_1)\ P_{x2} \\ x_1 &= R_{0,2}(\xi_2)\ P_{x0} + R_{1,2}(\xi_2)\ P_{x1} + R_{2,2}(\xi_2)\ P_{x2} \\ x_2 &= R_{0,2}(\xi_3)\ P_{x0} + R_{1,2}(\xi_3)\ P_{x1} + R_{2,2}(\xi_3)\ P_{x2} \\ y_0 &= R_{0,2}(\xi_1)\ P_{y0} + R_{1,2}(\xi_1)\ P_{y1} + R_{2,2}(\xi_1)\ P_{y2} \\ y_1 &= R_{0,2}(\xi_2)\ P_{y0} + R_{1,2}(\xi_2)\ P_{y1} + R_{2,2}(\xi_2)\ P_{y2} \\ y_2 &= R_{0,2}(\xi_3)\ P_{y0} + R_{1,2}(\xi_3)\ P_{y1} + R_{2,2}(\xi_3)\ P_{y2} \end{aligned} \tag{1.9}$$

In the matrix form

$$\mathbf{x} = \mathbf{T}\ \mathbf{P} \tag{1.10}$$

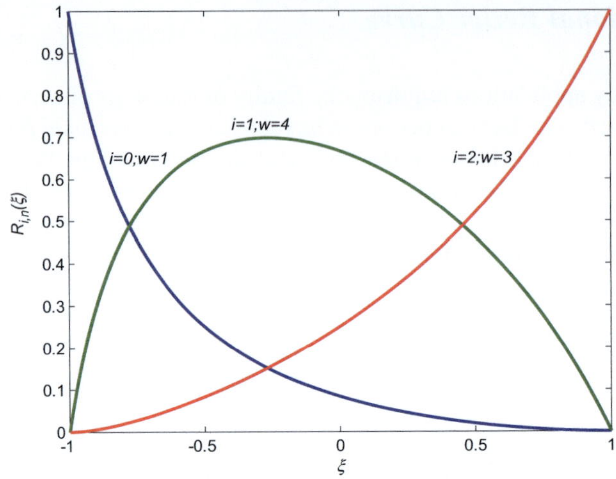

Fig. 1.9 Second order of rational Bézier basis functions ($n = 2$)

where

$$
\begin{aligned}
\mathbf{x} &= \begin{bmatrix} x_0 & x_1 & x_2 & y_0 & y_1 & y_2 \end{bmatrix}^T \\
\mathbf{P} &= \begin{bmatrix} P_{x0} & P_{x1} & P_{x2} & P_{y0} & P_{y1} & P_{y2} \end{bmatrix}^T \\
\mathbf{T} &= \begin{bmatrix}
R_{0,2}(\xi_1) & R_{1,2}(\xi_1) & R_{2,2}(\xi_1) & 0 & 0 & 0 \\
R_{0,2}(\xi_2) & R_{1,2}(\xi_2) & R_{2,2}(\xi_2) & 0 & 0 & 0 \\
R_{0,2}(\xi_3) & R_{1,2}(\xi_3) & R_{2,2}(\xi_3) & 0 & 0 & 0 \\
0 & 0 & 0 & R_{0,2}(\xi_1) & R_{1,2}(\xi_1) & R_{2,2}(\xi_1) \\
0 & 0 & 0 & R_{0,2}(\xi_2) & R_{1,2}(\xi_2) & R_{2,2}(\xi_2) \\
0 & 0 & 0 & R_{0,2}(\xi_3) & R_{1,2}(\xi_3) & R_{2,2}(\xi_3)
\end{bmatrix}
\end{aligned}
$$

MATLAB code RationalBezierBasis.m and function Bernstein.m (Sect. 1.2.2) draws the rational Bézier basis functions for $n = 2$ as shown in Fig. 1.9 which is based on the (1.8). Notice that we can have more variation of polynomial forms to compose a rational Bézier curve by changing the values of weights, w_i.

1.2.8 RationalBezierBasis Program List

```
% RationalBezierBasis.m
% Drawing the weighted Rational Bezier basis functions
% n = order of polynomial
% t = -1 to +1, incremented by dt
n=2;  % 2nd order polynomial
dt=1/50;
```

```
t=-1:dt:1;
nt=size(t,2);
wt=[1 4 3];
rb=zeros(nt,1);
for i=0:n
  for dt=1:nt
    Brnbasis = Bernstein(n,i,t(dt));
    rb(dt)=Brnbasis*wt(i+1);
    ydenom=0;
    for j=0:n
      Brnbasis = Bernstein(n,j,t(dt));
      ydenom=ydenom+Brnbasis*wt(j+1);
    end
    rb(dt)=rb(dt)/ydenom;
  end
  plot(t',rb,'LineWidth',2);
  hold all;
end
xlabel('\it{\xi}'); ylabel('\itR_{i,n}({\xi})');
text(-0.85,0.65,'\iti=0;w=1');
text(-0.4,0.72,'\iti=1;w=4');
text(0.45,0.65,'\iti=2;w=3');
```

1.2.9 RationalBezierCurve Program List

MATLAB code RationalBezierCurve.m and function Bernstein.m (Sect. 1.2.2) draw the rational Bézier curves for $n = 2$ as shown in Fig. 1.10 which is based on the (1.8). Notice that we have a different curve by changing the value of weights used in the rational Bézier curves rule.

```
% RationalBezierCurve.m
% Drawing the rational Bezier curve
% n = order of Bezier curve
n=2;
% Given nodal coordinates: xi and yi
xi=[-5 3 9];
yi=[ 2 5 2];
xyi=transpose([xi yi]);
% Parameter t
t=[-1 0 1];
```

```
w=[ 1 1 1]; % weights of x and y are equals
nt=size(t,2);
rmi=zeros(n+1,nt);
rdenom=zeros(nt,1);
for i=0:n
  for it=1:nt
    rdenom(it,1)=rdenom(it,1)+Bernstein(n,i,t(it))*w(i+1);
  end
end
for i=0:n
  for it=1:nt
    rmi(it,i+1) = Bernstein(n,i,t(it))*w(i+1)/rdenom(it,1);
  end
end
% [tm] matrix
tm=[rmi zeros(3,3);zeros(3,3) rmi];
% Control Points Px's and Py's calculation
pxy=tm\xyi;
px0=[pxy(1:nt)]; % n+1 control points
py0=[pxy(nt+1:2*nt)]; % n+1 control points
px=px0; py=py0;
wt=[ 1 4 3]; % n+1 weight x and y are equals
dt=1/40; % interval
t=-1:dt:1;
nt=size(t');
curvx=zeros(nt);
curvy=zeros(nt);
curvxR=zeros(nt);
curvyR=zeros(nt);
rdenom=zeros(nt);
for i=0:n
  for it=1:nt
    curvx(it)=curvx(it)+Bernstein(n,i,t(it))*px(i+1);
    curvy(it)=curvy(it)+Bernstein(n,i,t(it))*py(i+1);
    rdenom(it)=rdenom(it)+Bernstein(n,i,t(it))*wt(i+1);
  end
end
for i=0:n
  for it=1:nt
      curvxR(it)=curvxR(it)+Bernstein(n,i,t(it))*wt(i+1)*px(i+1)/
rdenom(it);
      curvyR(it)=curvyR(it)+Bernstein(n,i,t(it))*wt(i+1)*py(i+1)/
rdenom(it);
  end
end
```

```
plot(curvx,curvy,'-k','LineWidth',2);hold all;
plot(curvxR,curvyR,'-b','LineWidth',2);hold all;
axis([-6 10 1 9]);hold all;
plot(px,py,'--ko','LineWidth',1,'MarkerEdgeColor',...
           'k','MarkerFaceColor','w','MarkerSize',8);
hold all;
xlabel('\itx'); ylabel('\ity');
text(px(1),py(1)-0.5,'\itP_{0}');
text(px(2),py(2)+0.5,'\itP_{1}');
text(px(3),py(3)-0.5,'\itP_{2}');
text(2.8,6.4,'\itw=[1 4 3]');
text(2.8,5.2,'\itw=[1 1 1]');
```

1.2.10 What Is a Weight?

As we can see in Fig. 1.10 by using higher weight values more than one, we can increase the "*influence*" of the corresponding control points. The influence can be noticed from the changing curvature of the curve approaching the coordinates of the control points that are "*pulling*" the curve toward the control point. The higher the

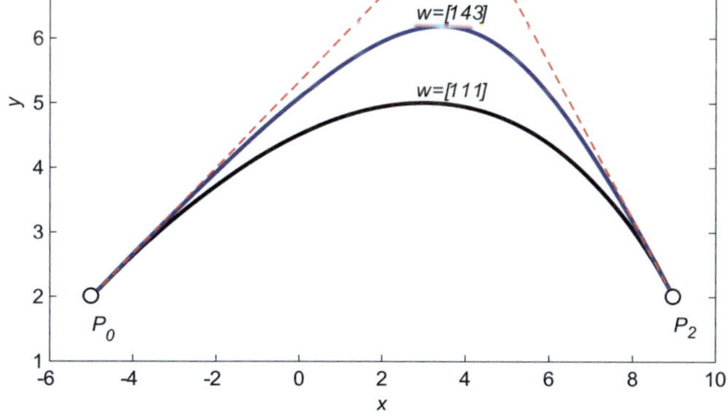

Fig. 1.10 A second-order rational Bézier curve constructed by given coordinates of three points (P_0, P_1, P_2), parametric vectors $\Xi = [-1 \quad 0 \quad 1]^T$ with weights $\mathbf{w} = [1 \quad 1 \quad 1]^T$ and $\mathbf{w} = [1 \quad 4 \quad 3]^T$

weight is given to the corresponding control point, the closer the curve to the point will be.

In practice, some curves such as circle, ellipse, and hyperbola cannot be constructed by using a simple Bézier curve; however, with the help of weighting values, those kinds of curves can be created.

Now, let us draw a circular curve by using the rational Bézier program given above. A circular curve can be drawn by using a second order ($n = 2$) of rational Bézier curve. We need three sampling points along the circular curve. Three sampling points were taken from both extremities of the arc and one at the middle $\Xi = [-1 \quad 0 \quad 1]^T$ of the curve. A simple trigonometry calculation can be performed to determine the coordinates. The uniform weights $\mathbf{w} = \left[1 \quad \cos\left(135^\circ/2\right) \quad 1\right]^T$ of each control point are used to calculate the coordinates of the control points as defined by (1.10). In fact, the second value of the weight vector is the cosines of half angle of the middle sampling point which can be derived from the quadratic, rational Bézier representation formula.

MATLAB code RationalBezierCurve.m (Sect. 1.2.9) and function Bernstein.m (Sect. 1.2.2) can draw the rational Bézier curves for $n = 2$ as shown in Fig. 1.11 which is based on the (1.8). Notice that we can obtain a circular curve by using the middle sampling point. For a 135° circular curve with a radius of 5 units, we can get the coordinates of the control points as $CP_0(-3.5355, 3.5355)$,

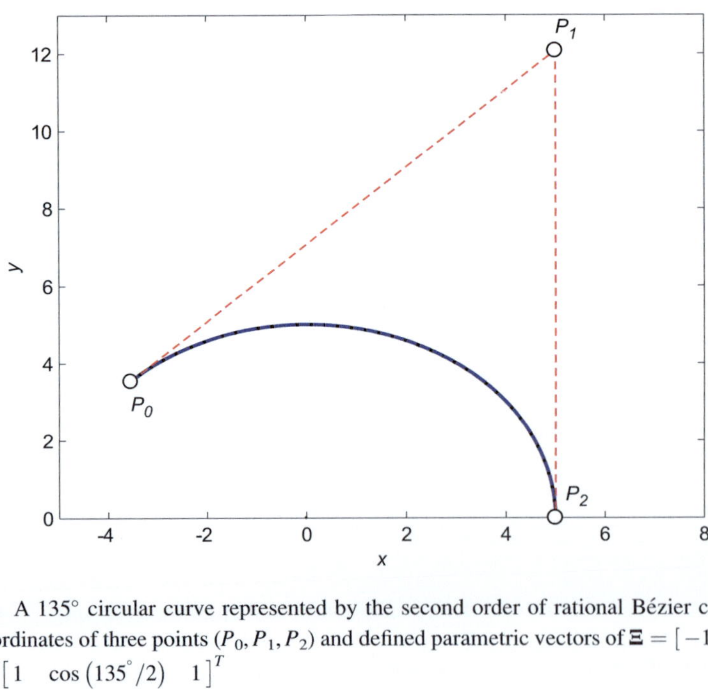

Fig. 1.11 A 135° circular curve represented by the second order of rational Bézier curve from given coordinates of three points (P_0, P_1, P_2) and defined parametric vectors of $\Xi = [-1 \quad 0 \quad 1]^T$ and $\mathbf{w} = \left[1 \quad \cos\left(135^\circ/2\right) \quad 1\right]^T$

$CP_1(5, 12.0711)$, and$CP_2(5, 0)$, which show the same results with the trigonometry calculation.

If we observe the circular curve carefully, in fact, the second control point was obtained from the intersection lines of both tangential directions at the extremity points. Then, we are in trouble if we try to construct a 180° of the circular curve since the computed second weight value will become zero. Hence, the intersection of both tangential lines from the extremities will intersect at the infinity. Approximately, to avoid instability in computing the infinity, we can set the value of the second weight to a very small value such as 0.0001. Thus we will get the coordinate of the second control point at $CP_1(0, 50000)$ which can be assumed to be infinite. This is one among the disadvantages of the rational Bézier curve because we need two curves of 90° of arc to represent a 180° arc.

1.2.11 Why Do We Need to Rationalize?

There are many necessities to have a curve being rationalized for practical usages. It is easier to understand the word *rationalization* as *normalization*. In organizing data from experiments, we need to normalize the data when seeking for the relationships between data within a particular range. The information of a relationship between one data compared to other data has become an important factor for generalization of a problem. The similarity between two or more data can be easily studied when the relationships among the data are related to each other. In practical data measurement applications, input data may need to be normalized to a specific lower range to prevent the ill-condition of the measurement machinery.

Normalization can also be used when we correlate different types of features, like comparing categorical vs. numeric, or date/time vs. numeric, then we must normalize the data before correlating them or search alternative methods of physical relationships.

In the upcoming chapter regarding the finite element analysis, we will deal with the so-called shape functions. The shape function is an important factor to be considered in the finite element method that leads to the accuracy and efficiency of the formulations and analyses. The shape function for the finite element formulation has to satisfy the partition of unity. These shape functions will be used for constructing the element stiffness matrix, mass matrix, loading vector, and other properties of a beam. Therefore, this inherited property by the Bézier curve will satisfy some requirements automatically in order of the curve being adopted as the shape function.

1.3 B-spline Curve

Very often we want to represent a curve that consists of more than one polynomial or rational segments. Numerically, we can do this by using a series of basis functions which can have different polynomials.

The pth-order of B-spline curve is defined by

$$\begin{aligned} x &= \sum_{i=0}^{n} N_{i,p}(\xi)\ P_{xi} \\ y &= \sum_{i=0}^{n} N_{i,p}(\xi)\ P_{yi} \end{aligned} \qquad -1 \le \xi \le 1 \tag{1.11}$$

In (1.11), $N_{i,p}(\xi)$ is the parametric ξ-based basis function (De Boor 1978, 1987). The B-spline polynomials are defined by

$$\begin{aligned} N_{i,0}(\xi) &= \begin{cases} 1 & \text{if } \xi_i \le \xi < \xi_{i+1} \\ 0 & \text{otherwise} \end{cases} \\ N_{i,p}(\xi) &= \frac{\xi - \xi_i}{\xi_{i+p} - \xi_i} N_{i,p-1}(\xi) + \frac{\xi_{i+p+1} - t}{\xi_{i+p+1} - \xi_{i+1}} N_{i+1,p-1}(\xi) \end{aligned} \tag{1.12}$$

The $N_{i,p}(\xi)$ are the pth degree B-spline basis functions defined on the nonperiodic knot vector Ξ. The knot vector Ξ consists of the parameters ξs in the range of $-1 \le \xi \le +1$. The ξs in (1.12) with subscript notation are fixed values that belong to the knot vector Ξ, while the ξs without any subscript notation are the variables.

1.3.1 BsplineBasis Program List

MATLAB code BsplineBasis.m and function Bspline.m (Sect. 1.3.2) draw the B-spline basis curves for $p = 3$ as shown in Fig. 1.12 which are based on the (1.11). Let us add an additional zero value inside the parametric vector so that it becomes $\Xi = [-1 \quad -1 \quad -1 \quad -1 \quad 0 \quad 1 \quad 1 \quad 1 \quad 1]^T$. The B-spline basis functions of the curves now become five as shown in Fig. 1.12. We can learn that by increasing the amount of parametric vector, the basis function will increase accordingly.

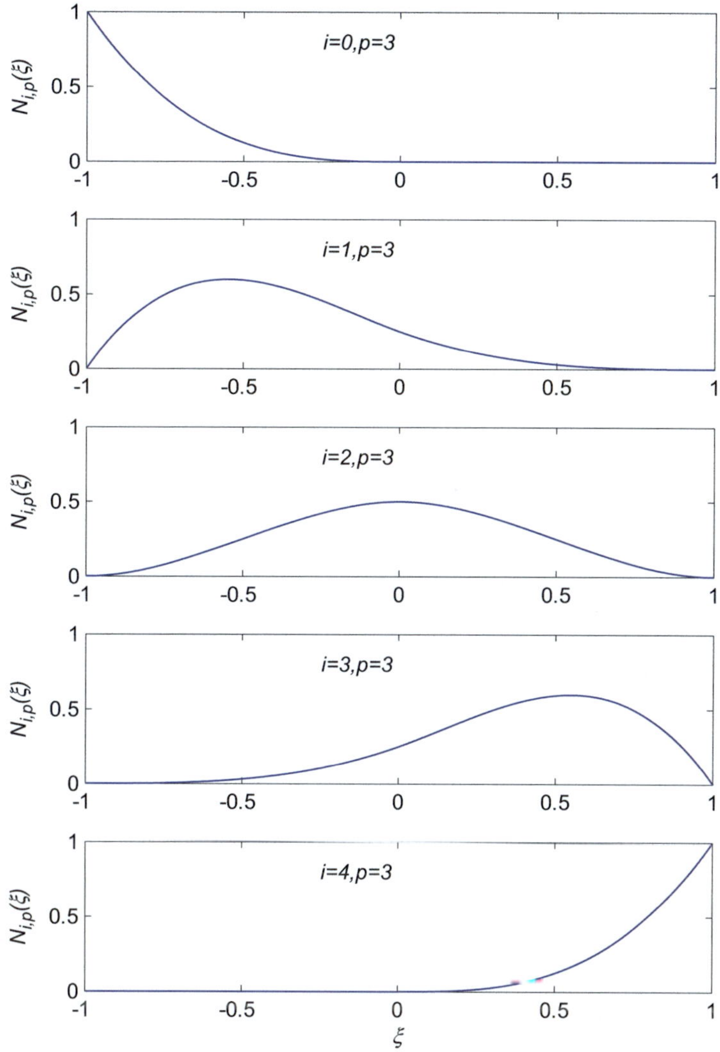

Fig. 1.12 Third-order of B-spline basis functions ($i = 0, 1, 2, 3, 4\ ; p = 3$) with a knot vector $\mathbf{\Xi} = [-1 \quad -1 \quad -1 \quad -1 \quad 0 \quad 1 \quad 1 \quad 1 \quad 1]^T$

```
% BsplineBasis.m
% Drawing the B-spline basis functions
% p = order of B-spline curves
clear;clc;
p=3;
knot=[-1 -1 -1 -1 0 1 1 1 1];
knot=[knot ones(1,p)];
```

```
nb=size(knot,2)-p-1;              % max number of basis functions
ndt=50;                           % t steps
t1=min(knot); t2=max(knot);
t=t1:(t2-t1)/ndt:t2;
nt=size(t,2);
y=zeros(nb,p+1,nt);               % BsplineBasisPlot (order,basis)
for it=1:nt
   Bspln = Bspline(p,knot,t(it));
   y(1:nb,1:(p+1),it) = Bspln(1:nb,1:(p+1));
end
for ip=0:p
 figure(ip+1);
 for ib=1:(nb-ip)
   subplot(nb-ip,1,ib);
   yC(1:nt)=y(ib,ip+1,1:nt);
   plot(t,yC,'LineWidth',2);
 end
end
xlabel('\it{\xi}');
```

1.3.2 Bspline Function List

```
function Bspln = Bspline(p,knot,t)
% Bspline.m
% p = order of B-spline curve
% knot = knot vector
% t = parameter
%---------------------------------------------------------------
nb=size(knot,2)-p-1;          % number of basis functions
Bspln=zeros(nb,p+1);          % B spline (order,basis)
for ip=0:p                    % order of polynomial
 for i=1:(nb-ip)              % i-th basis function
  iknot=knot(i);
  jknot=knot(i+1);
  kknot=knot(i+ip);
  lknot=knot(i+ip+1);
  if ip==0                    % ip = 0 order
    if t>=iknot && t<jknot
      Bspln(i,ip+1)=1;        % unity
    else
      Bspln(i,ip+1)=0;        % zero
    end
```

```
        if i==(nb-p) && t==jknot
          Bspln(i,ip+1)=1;         % unity at the right end to close
        end
      end
      if ip>0                      % ip > 0 order
        if kknot==iknot
           fract1=0;
        else
           fract1=(t-iknot)/(kknot-iknot);
        end
        if lknot==jknot
           fract2=0;
        else
           fract2=(lknot-t)/(lknot-jknot);
        end
           Bspln(i,ip+1)=fract1*Bspln(i,ip)+fract2*Bspln(i+1,ip);
      end
    end
  end
  return
```

Similar with the previous section, suppose we want to obtain the control points $P_{x0}, P_{x1}, P_{x2}, P_{y0}, P_{y1}$ and P_{y2} to represent a second-order polynomial by using the B-spline curve, we need the coordinates of three points of the curve. To do this, we use the same coordinates (x,y) of the three points (P_0, P_1, and P_2) in Sect. 1.1 and substituting into (1.4) results in

$$
\begin{aligned}
x_0 &= N_{0,2}(\xi_1)\ P_{x0} + N_{1,2}(\xi_1)\ P_{x1} + N_{2,2}(\xi_1)\ P_{x2} \\
x_1 &= N_{0,2}(\xi_2)\ P_{x0} + N_{1,2}(\xi_2)\ P_{x1} + N_{2,2}(\xi_2)\ P_{x2} \\
x_2 &= N_{0,2}(\xi_3)\ P_{x0} + N_{1,2}(\xi_3)\ P_{x1} + N_{2,2}(\xi_3)\ P_{x2} \\
y_0 &= N_{0,2}(\xi_1)\ P_{y0} + N_{1,2}(\xi_1)\ P_{y1} + N_{2,2}(\xi_1)\ P_{y2} \\
y_1 &= N_{0,2}(\xi_2)\ P_{y0} + N_{1,2}(\xi_2)\ P_{y1} + N_{2,2}(\xi_2)\ P_{y2} \\
y_2 &= N_{0,2}(\xi_3)\ P_{y0} + N_{1,2}(\xi_3)\ P_{y1} + N_{2,2}(\xi_3)\ P_{y2}
\end{aligned}
\tag{1.13}
$$

In the matrix form

$$
\mathbf{x} = \mathbf{T}\ \mathbf{P} \tag{1.14}
$$

where

$$\mathbf{x} = \begin{bmatrix} x_0 & x_1 & x_2 & y_0 & y_1 & y_2 \end{bmatrix}^T$$
$$\mathbf{P} = \begin{bmatrix} P_{x0} & P_{x1} & P_{x2} & P_{y0} & P_{y1} & P_{y2} \end{bmatrix}^T$$
$$\mathbf{T} = \begin{bmatrix} N_{0,2}(\xi_1) & N_{1,2}(\xi_1) & N_{2,2}(\xi_1) & 0 & 0 & 0 \\ N_{0,2}(\xi_2) & N_{1,2}(\xi_2) & N_{2,2}(\xi_2) & 0 & 0 & 0 \\ N_{0,2}(\xi_3) & N_{1,2}(\xi_2) & N_{2,2}(\xi_3) & 0 & 0 & 0 \\ 0 & 0 & 0 & N_{0,2}(\xi_1) & N_{1,2}(\xi_1) & N_{2,2}(\xi_1) \\ 0 & 0 & 0 & N_{0,2}(\xi_2) & N_{1,2}(\xi_2) & N_{2,2}(\xi_2) \\ 0 & 0 & 0 & N_{0,2}(\xi_3) & N_{1,2}(\xi_3) & N_{2,2}(\xi_3) \end{bmatrix}$$

MATLAB code BsplineCurve.m uses function Bspline.m (Sect. 1.3.2) and solves the values of the vector **P**, further constructs the B-spline polynomial equation in (1.11), and finally draws the B-spline curve which is resulted in the same curve as shown in Fig. 1.6.

1.3.3 BsplineCurve Program List

```
% BsplineCurve.m
% Drawing a 2nd order Bspline curve from a given three points
% p = order of Bspline curves
clear all;clc;
p=2;
% Nodal coordinates: xi and yi
xi=[-5 3 9];
yi=[ 2 5 2];
xyi=transpose([xi yi]);
% Knot vector
knot=[-1 -1 -1  1  1  1];
knot=[knot ones(1,p)];
nb=size(knot,2)-p-1;          % max number of basis functions
% Parameter t
t=[-1 0 1];
nt=size(t,2);
tmi=zeros(nt,p+1);
for i=0:p
  for dt=1:nt
    Bspln = Bspline(p,knot,t(dt));
    tmi(dt,i+1) = Bspln(i+1,p+1);
  end
end
```

```
% [tm] matrix
tm=[tmi zeros(3,3);zeros(3,3) tmi];
% Coefficients Px's and Py's calculation
pxy=tm\xyi;
px0=[pxy(1:nt)];       % n+1 control points
py0=[pxy(nt+1:2*nt)]; % n+1 control points
px=px0; py=py0;
%-----------------------------------------------------------
ndt=100;                    % t steps
t1=min(knot); t2=max(knot);
t=t1:(t2-t1)/ndt:t2;
nt=size(t,2);
curvx=zeros(nt,1);          % BsplineCurve variables
curvy=zeros(nt,1);
for ib=1:nb-p               % number of basis function
  for it=1:nt               % number of step
      Bspln = Bspline(p,knot,t(it));
      curvx(it)=curvx(it)+Bspln(ib,p+1)*px(ib);
      curvy(it)=curvy(it)+Bspln(ib,p+1)*py(ib);
  end
end
plot(curvx,curvy,'-k','LineWidth',2);
hold all;
plot(xi,yi,'ko','LineWidth',1,'MarkerEdgeColor',...
           'k','MarkerFaceColor','w','MarkerSize',8);
plot(px,py,'--ro','LineWidth',1,'MarkerEdgeColor',...
           'k','MarkerFaceColor','r','MarkerSize',8);
hold all;
axis([-6 10 1 9]);
text(px(1)-0.5,py(1)+0.5,'CP0');
text(px(2),py(2)+0.5,'CP1');
text(px(3),py(3)+0.5,'CP2');
text(xi(1),yi(1)-0.5,'P0');
text(xi(2),yi(2)+0.5,'P1');
text(xi(3),yi(3)-0.5,'P2');
xlabel('X'); ylabel('Y');
```

The coordinates of the control points which were obtained from the calculation are $CP_0(-5, 2)$, $CP_1(4, 8)$, and $CP_2(9, 2)$.

Note that the B-spline curves are suitable for intuitive design by using free-form curves. This is because the knot vector gives the designer more handles to harness the curve creations. Now, let us explore the capability of B-spline polynomial in creating fascinating curves that the Bézier polynomial functions cannot do.

To demonstrate the capability of B-spline formulation in combining two polynomial forms into one formulation, the examples shown in Fig. 1.8 will be used.

MATLAB code RationalBsplineCP.m uses the function Bspline.m (Sect. 1.3.2) to construct B-spline curves by using a set of control points. Two polynomial forms in the B-spline curve are representations by one B-spline formulation as shown in Fig. 1.13.

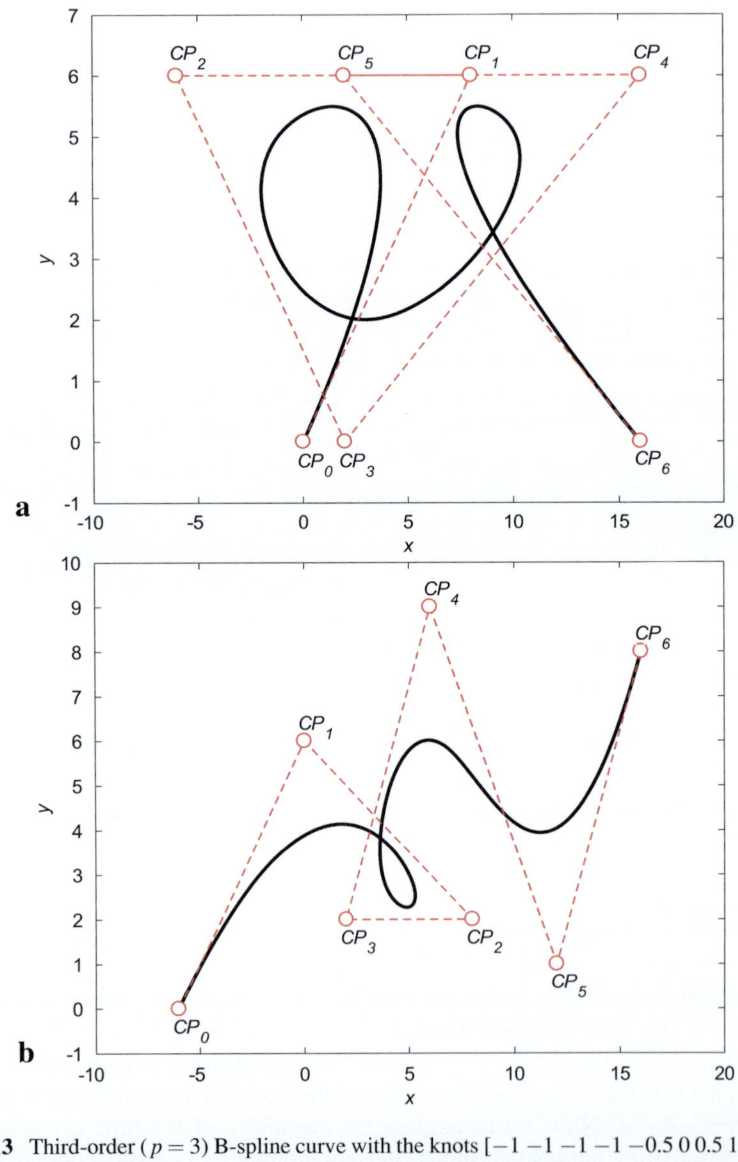

Fig. 1.13 Third-order ($p = 3$) B-spline curve with the knots $[-1\ -1\ -1\ -1\ -0.5\ 0\ 0.5\ 1\ 1\ 1\ 1]$ by joining two polynomial curve examples shown in Fig. 1.8 with the following control point data: (**a**) Loop and Cusp [(0,0); (8,6); (−6,6); (2,0); (16,6); (2,6); (16,0)], (**b**) Hook and Inflection [(−6,0); (0,6); (8,2); (2,2); (6,9); (12,1); (16,8)]

1.3.4 BsplineCurveCP Program List

```
% BsplineCurveCP.m
% Drawing the Bspline curve using Control Points
% p = order of Bspline curves
clear all;clc;
p=3;
knot=[-1 -1 -1 -1 -0.5 0 0.5 1 1 1 1]; % uniform interval
px0=[ 0 8 -6 2 16  2 16];             % n+1 control points
py0=[ 0 6  6 0  6  6  0];             % n+1 control points
knot=[knot ones(1,p)];
px=px0; py=py0;
nb=size(knot,2)-p-1;                   % max number of basis
                                         functions
ndt=100;                               % steps of parameter t
t1=min(knot); t2=max(knot);
t=t1:(t2-t1)/ndt:t2;
nt=size(t,2);
curvx=zeros(nt,1);                     % BsplineCurve variables
curvy=zeros(nt,1);
for ib=1:nb-p                          % number of basis function
  for it=1:nt
      Bspln = Bspline(p,knot,t(it));
      curvx(it)=curvx(it)+Bspln(ib,p+1)*px(ib);
      curvy(it)=curvy(it)+Bspln(ib,p+1)*py(ib);
  end
end
plot(curvx,curvy,'-k','LineWidth',2);
hold all;
plot(px,py,'--ro','LineWidth',1,'MarkerEdgeColor',...
          'r','MarkerFaceColor','w','MarkerSize',8),
hold all;
axis([-10 20 -1 7]);
text(px(1)-0.3,py(1)-0.3,'\itCP_0');
text(px(2)-0.3,py(2)+0.3,'\itCP_1');
text(px(3)-0.3,py(3)+0.3,'\itCP_2');
text(px(4)-0.3,py(4)-0.3,'\itCP_3');
text(px(5)-0.3,py(5)+0.3,'\itCP_4');
text(px(6)-0.3,py(6)+0.3,'\itCP_5');
text(px(7)-0.3,py(7)-0.3,'\itCP_6');
xlabel('X'); ylabel('Y');
```

We can see in the figures two polynomial curves are combined into one system of polynomial in B-spline curve formulation. However, at the connecting point, we observed also the curves changed in order to maintain a certain level the continuity that will be discussed in the following section.

1.3.5 The Roles of Knot Vector

A typically known vector is given by the following equation:

$$\Xi = \begin{bmatrix} \underbrace{-1 \quad . \quad . \quad -1}_{p\ \ knots} & \underbrace{-1 \quad (\xi \ldots)_{m(i)} \quad . \quad . \quad (\xi \ldots)_{m(i+1)} \quad 1}_{knots} & \underbrace{1 \quad . \quad . \quad 1}_{p\ \ knots} \end{bmatrix}^T \tag{1.15}$$

where m_i is a repeated m value of knots in the range of $[-1 \quad 1]$. The continuity condition of the curve at the repetitive knots can be defined as

$$C^{p-m} \tag{1.16}$$

Various orders of parametric level of continuity between curves can be described as follows:

- C^{-1}: the curves have discontinuities.
- C^0: the curves are connected to each other.
- C^1: the first derivative of the curves are continuous.
- C^2: the 1st and 2nd derivatives of the curves are continuous.
- C^n: the 1st through nth derivatives of the curves are continuous.

As we can see the knot vector in (1.15), each knot can be considered as a *breakpoint* of several polynomial curves. Therefore, one polynomial will be represented between two non-repeating knots. For a $p = 2$ polynomial curve, if the two of three control points are to be fixed at the extremities of the curve, the first and the last knots have to be repeated $p + 1$ times as the knots. This will enforce a level of discontinuity C^{-1} at both extremity points. Hence, both points will be "*clamped*" at both extremities of the curve.

The simplest knot vector of the $p = 2$ polynomial B-spline curve representation can be given as

$$\Xi = \begin{bmatrix} \underbrace{-1 \quad -1}_{p=2\ \ knots} & \underbrace{-1 \qquad 1}_{3\ \ basis\ \ functions} & \underbrace{1 \quad 1}_{p=2\ \ knots} \end{bmatrix}^T \tag{1.17}$$

From Eq. (1.17), we can observe there are six knots; but because three knots are repetitive two-number knots, essentially only two knots are unique. The first and the last p basis functions will not generate a basis function. The control points required within the range of $[-1 \quad 1]$ are three. Because of this, we have three corresponding basis functions which are exactly similar to the Bernstein polynomial basis functions as shown in Fig. 1.7.

Implicitly, by inserting a knot inside the interval, we increase the number of basis functions, and it also means that we have to increase the amount of control points required to draw a B-spline curve without increasing the order of polynomial of the curve.

1.3.6 Rational B-Spline Curve

The nonuniform rational B-spline (NURBS) requires the formulation of rationalized B-spline. Rationalizing the B-spline is formulated in the same manner with the other curves previously discussed.

The nth order of rational B-spline curve is defined by

$$\begin{aligned} x &= \sum_{i=0}^{n} S_{i,p}(\xi)\ P_{xi} \\ y &= \sum_{i=0}^{n} S_{i,p}(\xi)\ P_{yi} \end{aligned} \qquad -1 \leq \xi \leq 1 \tag{1.18}$$

where

$$S_{i,p}(\xi) = \frac{N_{i,p}(\xi)\ w_i}{\sum_{j=0}^{n} N_{j,p}(\xi)\ w_j} \qquad w_i > 0$$

The w_i are *weights* of control points. The $S_{i,p}(\xi)$ are the rational basis functions for the B-spline curve.

Now, let us see if we use the same sampling point data from Sect. 1.1 with the weights of $w_0 = 1$, $w_1 = 2$, and $w_2 = 6$.

MATLAB code RationalBspline.m uses the function Bspline.m (Sect. 1.3.2) and solves the values of control point vector $\mathbf{P}$ and further constructs the rationalized B-spline basis functions which are shown in Fig. 1.14.

1.3.7 RationalBsplineBasis Program List

```
% RationalBsplineBasis.m
% Drawing the rational Bspline basis functions
% p = order of Bspline curves
clear;clc;
p=2;
```

```
knot=[-1 -1 -1 1 1 1];                % Bezier Curve, 3 CPs
knot=[knot ones(1,p)];
nb=size(knot,2)-p-1;                  % max number of basis functions
ndt=100;                              % t steps
t1=min(knot); t2=max(knot);
t=t1:(t2-t1)/ndt:t2;
nt=size(t,2);
wt=[1 2 6]';                          % Non-Uniform weights
Sb=zeros(nt,1);                       % BsplineBasisPlot (order,basis)
for j=1:(nb-p)
  for it=1:nt
    Bspln = Bspline(p,knot,t(it));
    Sb(it,1)=Bspln(j,p+1)*wt(j);
    ydenom=0;
    for i=1:(nb-p)
      ydenom=ydenom+Bspln(i,p+1)*wt(i);
    end
    Sb(it,1)=Sb(it,1)/ydenom;
  end
  plot(t',Sb,'LineWidth',2);
  hold all;
end
xlabel('t'); ylabel('Si,p(t)');
```

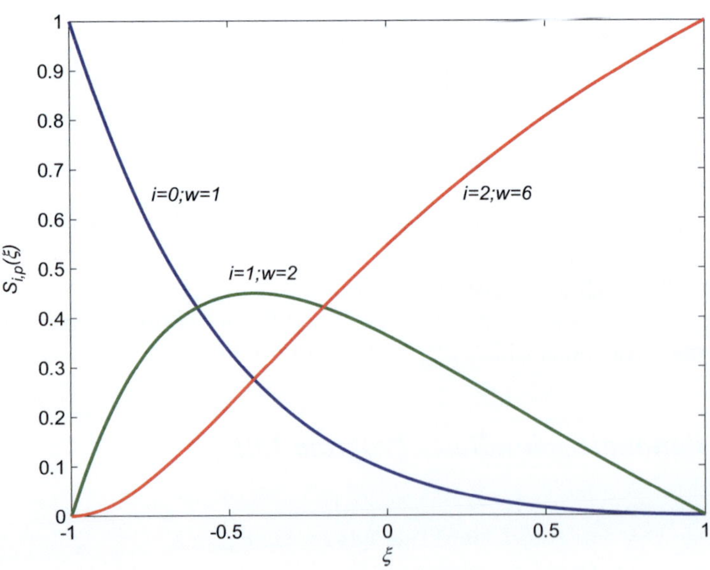

Fig. 1.14 Second-order of B-spline basis functions ($p=2$) with a knot vector $\boldsymbol{\Xi} = [-1 \quad -1 \quad -1 \quad 1 \quad 1 \quad 1]^T$ and weight vector $\mathbf{w} = [1 \quad 2 \quad 6]^T$

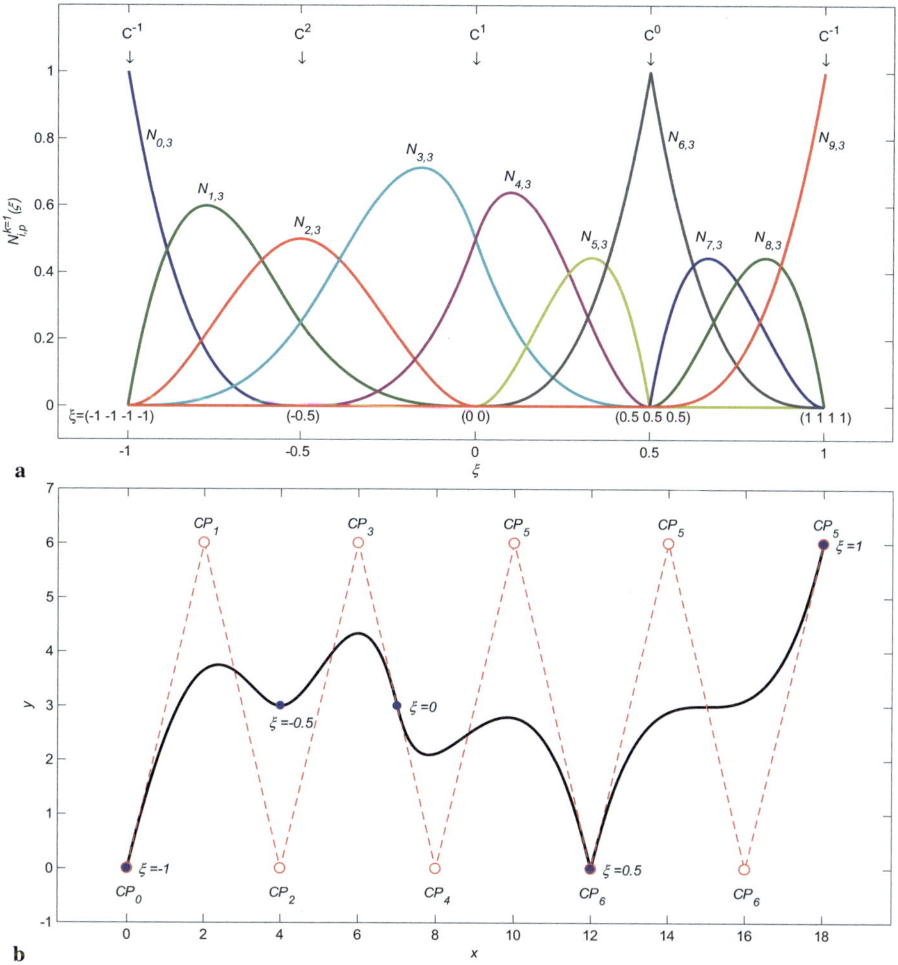

Fig. 1.15 (**a**) Third order of NURBS basis functions ($p=3$) with a knot vector $\Xi=[-1\quad -1\quad -1\quad -1\quad -0.5\quad 0\quad 0\quad 0.5\quad 0.5\quad 0.5\quad 1\quad 1\quad 1\quad 1]^T$ with uniform weight vector (**b**) The NURBS curve drawn using the following control points [(0,0); (2,6); (4,0); (6,6); (8,0); (10,6); (12,0), (14,6), (16,0), (18,6)]

1.3.8 NURBS Curve

The term NURBS is an abbreviation of nonuniform rational B-spline. NURBS is a particular case of B-spline curve where the knot vector is not uniform. The nonuniformity of the knot vector is necessary in order to draw a system of polynomial curves with the desired level of continuity (Sect. 1.3.5) at several knots.

MATLAB code NurbsBasis.m uses function Nurbs.m (Sect. 1.3.10) to draw NURBS basis functions which are shown in Fig. 1.15.

1.3.9 NURBSBasis Program List

```
% NurbsBasis.m
% Drawing the NURBS basis functions
% p = order of Bspline curves
clear;clc;
p=3;
knot=[-1 -1 -1 -1 -0.5 0 0 0.5 0.5 0.5 1 1 1 1]; %Nonuniform
Bspline
knot=[knot ones(1,p)];
nb=size(knot,2)-p-1;                % max number of basis functions
ndt=100;
t1=min(knot); t2=max(knot);
t=t1:(t2-t1)/ndt:t2;
nt=size(t,2);
wt=ones(nb-p);                      % Uniform weights
figure(2);
Sb=zeros(nt,1);                     % BsplineBasisPlot (order,basis)
for j=1:(nb-p)
  for it=1:nt
    Nurbsbasis = Nurbs(p,knot,wt,t(it));
    Sb(it,1)=Nurbsbasis(j,p+2);
  end
  plot(t',Sb,'LineWidth',2);
  hold all;
end
axis([-1.2 1.2 -0.1 1.2]); set(gca,'XTick',-1:0.5:1);
text(-1.18,-0.03,'knots:(-1 -1 -1 -1)');
text(-0.54,-0.03,'(-0.5)'); text(-0.04,-0.03,'(0 0)')
text(0.4,-0.03,'(0.5 0.5 0.5)'); text(0.93,-0.03,'(1 1 1 1)');
text(-1.02,1.12,'C^{-1}'); text(-1.01,1.04,'\downarrow');
text(-0.52,1.12,'C^{2}'); text(-0.51,1.04,'\downarrow');
text(-0.02,1.12,'C^{1}'); text(-0.01,1.04,'\downarrow');
text(0.48,1.12,'C^{0}'); text(0.49,1.04,'\downarrow');
text(0.98,1.12,'C^{-1}'); text(0.99,1.04,'\downarrow');
text(-0.95,0.8,'\itN_{0,3}'); text(0.55,0.8,'\itN_{6,3}');
text(0.98,0.8,'\itN_{9,3}'); text(-0.8,0.64,'\itN_{1,3}');
text(-0.52,0.54,'\itN_{2,3}'); text(-0.2,0.76,'\itN_{3,3}');
text(0.08,0.68,'\itN_{4,3}'); text(0.3,0.5,'\itN_{5,3}');
text(0.63,0.5,'\itN_{7,3}'); text(0.8,0.5,'\itN_{8,3}');
xlabel('\itt'); ylabel('\itN_{i,p}(\itt)');
```

1.3.10 Nurbs Function List

```
% Nurbs.m
function Nurbsbasis = Nurbs(p,knot,wt,t)
% Nurbs.m
% p = order of polynomial
% knot = knot vector
% w = weight vector
% t = parameter
%-----------------------------------------------------------------
nb=size(knot,2)-p-1;          % number of basis functions
% Constructing NURBS basis functions
Nurbsbasis=zeros(nb,p+1+1);  % NURBS (order,basis)
for ip=0:p                    % order of polynomial
 for i=1:(nb-ip)              % i-th basis function
  iknot=knot(i);
  jknot=knot(i+1);
  kknot=knot(i+ip);
  lknot=knot(i+ip+1);
  if ip==0                    % ip = 0 order
    Nurbsbasis(i,ip+1)=1;
    if t>=iknot && t<jknot
      Nurbsbasis(i,ip+1)=1;        % unity
    else
      Nurbsbasis(i,ip+1)=0;        % zero
    end
    if i==(nb-p) && t==jknot
      Nurbsbasis(i,ip+1)=1;        % unity
    end
  end
  if ip>0                     % ip > 0 order
    if kknot==iknot
     fract1=0;
    else
     fract1=(t-iknot)/(kknot-iknot);
    end
    if lknot==jknot
     fract2=0;
    else
     fract2=(lknot-t)/(lknot-jknot);
    end
     Nurbsbasis(i,ip+1)=fract1*Nurbsbasis(i,ip)+fract2*Nurbsbasis
(i+1,ip);
```

```
  end
 end
end
% Constructing Rational NURBS basis functions
for j=1:(nb-p)
    Nurbsbasis(j,p+2)=Nurbsbasis(j,p+1)*wt(j);
    denom=0;
    for i=1:(nb-p)
      denom=denom+Nurbsbasis(i,p+1)*wt(i);
    end
    Nurbsbasis(j,p+2)=Nurbsbasis(j,p+2)/denom;
end
return
```

To demonstrate the capability of NURBS in combining two polynomial forms into one formulation, the examples shown in Fig. 1.8 will be used.

MATLAB code NurbsCurveDrawCP.m uses the function Nurbs.m (Sect. 1.3.10) to construct the NURBS curve by using the same control points taken from the examples in Fig. 1.8. The presentations of two polynomial forms are shown in Fig. 1.16. We can see the capability of the NURBS to connect two polynomial curves into a single formulation. By inserting $[.\ \ .\ \ .\ \ 0\ \ 0\ \ 0\ \ 0\ \ .\ \ .\ \ .]$ knots into the Ξ vector, it means we can get a $C^{p-m=3-3=0}$ continuity condition that connects both control points of the two polynomials.

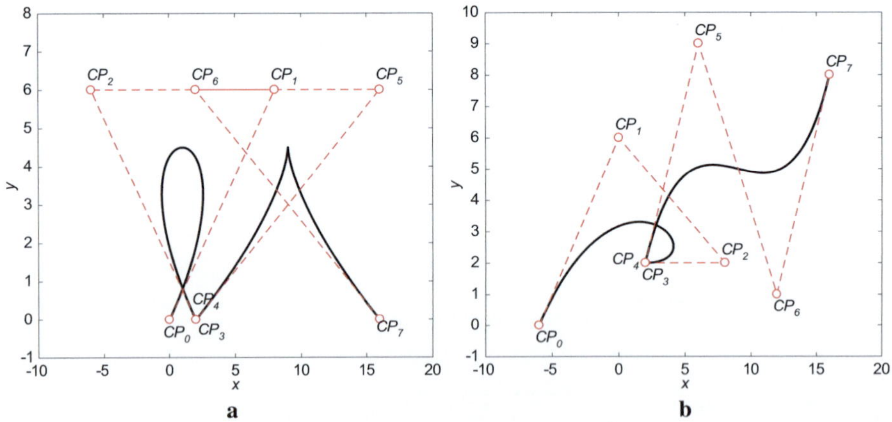

Fig. 1.16 Third order of NURBS basis functions ($p = 3$) with a knot vector $\Xi = [-1\ \ -1\ \ -1\ \ -1\ \ 0\ \ 0\ \ 0\ \ 0\ \ 1\ \ 1\ \ 1\ \ 1]^T$ and uniform weight vector with the following control point data: (**a**) Loop and Cusp [(0,0); (8,6); (−6,6); (2,0); (2,0); (16,6); (2,6); (16,0)], (**b**) Hook and Inflection [(−6,0); (0,6); (8,2); (2,2); (6,9); (12,1); (16,8)]

1.3.11 NURBSCurveDrawCP Program List

```
% NurbsCurveDrawCP.m
% Drawing the NURBS curve using Control Points
% p = order of Bspline curves
clear all;clc;
p=3;
knot=[-1 -1 -1 -1 0 0 0 0 1 1 1 1]; % uniform interval
px0=[ 0 8 -6 2 2 16  2 16];          % Fig. 1.9 a and b
py0=[ 0 6  6 0 0 6  6  0];           %
%px0=[-6 0 8 2 2 6 12 16];           % Fig. 1.9 c and d
%py0=[ 0 6 2 2 2 9  1  8];           %
knot=[knot ones(1,p)];
px=px0; py=py0;
%-----------------------------------------------------------
nb=size(knot,2)-p-1;              % max number of basis functions
ndt=100;                          % steps of parameter t
t1=min(knot); t2=max(knot);
t=t1:(t2-t1)/ndt:t2;
nt=size(t,2);
wt=ones(nb-p);                    % Uniform weights
curvx=zeros(nt,1);                % Nurbs variables
curvy=zeros(nt,1);
for ib=1:nb-p                     % Number of basis function
  for it=1:nt
    Nurbsbasis = Nurbs(p,knot,wt,t(it));
    curvx(it)=curvx(it)+Nurbsbasis(ib,p+2)*px(ib);
    curvy(it)=curvy(it)+Nurbsbasis(ib,p+2)*py(ib);
  end
end
plot(curvx,curvy,'-k','LineWidth',2);
hold all;
plot(px,py,'--ro','LineWidth',1,'MarkerEdgeColor',...
          'r','MarkerFaceColor','w','MarkerSize',8);
hold all; axis([-10 20 -2 8]);
text(px(1)-0.3,py(1)-0.5,'\itCP_0');  text(px(2)-0.3,py(2)
+0.3,'\itCP_1');
text(px(3)-0.3,py(3)+0.3,'\itCP_2');  text(px(4)-0.3,py(4)-
0.5,'\itCP_3');
text(px(5)-0.3,py(5)+0.3,'\itCP_4');  text(px(6)-0.3,py(6)
+0.3,'\itCP_5');
text(px(7)-0.3,py(7)-0.5,'\itCP_6'); xlabel('X'); ylabel('Y');
```

1.3.12 Derivatives of NURBS Basis Function

Since the NURBS is constructed from the rational B-spline curves family, the derivative of the NURBS (DNURBS) also can be obtained from the derivative of the B-spline curves. The derivative of basis functions of the B-spline curves is given by

$$N_{i,p}^{k}(\xi) = \frac{p}{\xi_{\max} - \xi_{\min}} \left(\frac{N_{i,p-1}^{k-1}(\xi)}{\xi_{i+p} - \xi_i} - \frac{N_{i+1,p-1}^{k-1}(\xi)}{\xi_{i+p+1} - \xi_{i+1}} \right) \tag{1.19}$$

where the superscript k denotes the kth derivative of the basis function.

The derivative of the basis functions of the B-spline curves can be obtained by using the above equation multiplied by their corresponding weights, similar with (1.18). The derivatives of NURBS then can be constructed by rationalizing the derivatives of the basis functions of the B-spline above as follows:

$$S_{i,p}^{k}(\xi) = \frac{N_{i,p}^{k}(\xi)\ w_i}{\sum_{j=0}^{n} N_{i,p}^{k}(\xi)\ w_j} \qquad w_i > 0 \tag{1.20}$$

The w_i are *weights* of the control points. The $S_{i,p}^{k}(\xi)$ are the rational basis functions for the NURBS curve.

MATLAB code DNurbsBasis.m uses the function DNurbsLeibnitz.m (Sect. 1.3.14) to draw the first and second derivative of NURBS basis functions which are shown in Fig. 1.17.

1.3.13 DNURBSBasis Program List

```
% DNurbsBasis.m
% DNurbsBasis.m
% Drawing the 1st and 2nd derivatives of NURBS basis functions
% p = degree of Bspline curves
clear;clc;
p=3;
knot=[-1 -1 -1 -1 -0.5 0 0 0.5 0.5 0.5 1 1 1 1]; %Nonuniform
Bspline
knot=[knot ones(1,p)];
nb=size(knot,2)-p-1;                % max number of basis functions
ndt=50;                             % t steps
t1=min(knot); t2=max(knot);
t=t1:(t2-t1)/ndt:t2;
nt=size(t,2);
```

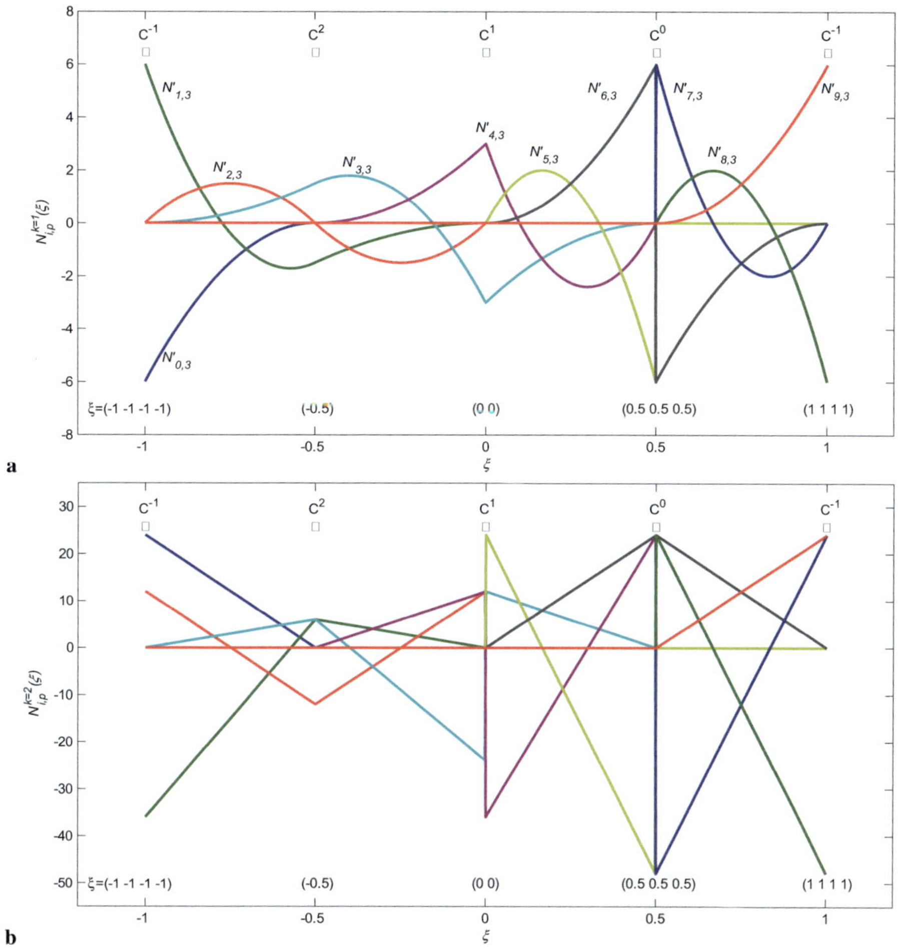

Fig. 1.17 (**a**) First ($k = 1$) and (**b**) second ($k = 2$) derivatives of the NURBS basis functions ($p = 3$) with a knot vector $\Xi = [-1 \quad -1 \quad -1 \quad -1 \quad -0.5 \quad 0 \quad 0 \quad 0.5 \quad 0.5 \quad 0.5 \quad 1 \quad 1 \quad 1 \quad 1]^T$ and unity weight vector

```
wt=ones(nb-p);                          % uniform weights
Sb=zeros(nt,1);
figure(2);
for j=1:(nb-p)
  for it=1:nt
    DNurbsbasis = DNurbsLeibnitz(p,knot,wt,t(it));
    Sb(it,1)=DNurbsbasis(j,p+7);  %1st derivative of the function
   %Sb(it,1)=DNurbsbasis(j,p+8);  %2nd derivative of the function
  end
  plot(t',Sb,'LineWidth',2);
  hold all;
end
```

```
axis([-1.2 1.2 -8 8]); set(gca,'XTick',-1:0.5:1);
text(-1.18,-7,' {\xi}=(-1 -1 -1 -1)');
text(-0.54,-7,'(-0.5)'); text(-0.04,-7,'(0 0)')
text(0.4,-7,'(0.5 0.5 0.5)'); text(0.93,-7,'(1 1 1 1)');
text(-1.02,7.5,'C^{-1}'); text(-1.01,6.5,'\downarrow');
text(-0.52,7.5,'C^{2}'); text(-0.51,6.5,'\downarrow');
text(-0.02,7.5,'C^{1}'); text(-0.01,6.5,'\downarrow');
text(0.48,7.5,'C^{0}'); text(0.49,6.5,'\downarrow');
text(0.98,7.5,'C^{-1}'); text(0.99,6.5,'\downarrow');
text(-0.95,5,'\itN"_{1,3}'); text(0.55,5,'\itN"_{7,3}');
text(0.98,5,'\itN"_{9,3}'); text(-0.8,2,'\itN"_{2,3}');
text(-0.95,-5.2,'\itN"_{0,3}'); text(-0.42,2.2,'\itN"_{3,3}');
text(-0.03,3.5,'\itN"_{4,3}'); text(0.13,2.6,'\itN"_{5,3}');
text(0.3,5,'\itN"_{6,3}'); text(0.65,2.6,'\itN"_{8,3}');
xlabel('\it{\xi}'); ylabel('\itN_{i,p}^{k=1}(\it{\xi})');
```

1.3.14 DNurbsLeibnitz Program List

```
function DNurbsbasis = DNurbsLeibnitz(p,knot,wt,t)
% DNurbsLeibnitz.m (R7) for Nonuniform knots
% p = degree of Bspline curves, knot = knot vector, t = parameter
% storage info: N_i^0 (1 to p+1); N_i' (p+2 to p+4)
% storage info: R_i^0 (p+5);    R_i' and R_i" (p+7 to p+8)
%-----------------------------------------------------------------
nb=size(knot,2)-p-1;                 % number of basis functions
% Constructing Bspline basis functions (p+1)
DNurbsbasis=zeros(nb,nb-p+11);       % B spline (order,basis)
for ip=0:p                           % degree of polynomial
 for i=1:(nb-ip)                     % number of basis function
  iknot=knot(i);
  jknot=knot(i+1);
  kknot=knot(i+ip);
  lknot=knot(i+ip+1);
  if ip==0
    if t>=iknot && t<jknot           % ip = 0 degree
      DNurbsbasis(i,ip+1)=1;         % unity
    else
      DNurbsbasis(i,ip+1)=0;         % zero
    end
    if i==(nb-p) && t==jknot
      DNurbsbasis(i,ip+1)=1;         % unity at the right end to
                                       close
    end
  end
```

```
  if ip>0                          % ip = 0 degree
    if kknot==iknot
       fract1=0;
    else
       fract1=(t-iknot)/(kknot-iknot);
    end
    if lknot==jknot
       fract2=0;
    else
       fract2=(lknot-t)/(lknot-jknot);
    end
       DNurbsbasis(i,ip+1)=fract1*DNurbsbasis(i,ip)+...
                         fract2*DNurbsbasis(i+1,ip);
  end
 end
end
% Constructing Bspline basis functions (p+2~2p+1)
% Derivative Basis Functions (only for the k's 1st and 2nd der the
p basis)
for k=1:3                        % 1st der for p-1,p and 2nd der
                                   for p only
  if k<3
     ip=p+k-2;                   % p=p-1
     npi=p+1+k-3;                % storage array
     if ip == 0
         ip=1;
         npi=1;
     end
  else
     ip=p;                       % p=p
     npi=p+1+1;                  % storage array
  end
  for i=1:nb-ip                  % corresponding basis function
                                   i=1:p+1
     k_i_knot=(knot(i+ip)-knot(i));
     m_j_knot=(knot(i+ip+1)-knot(i+1));
     if k_i_knot < 1.E-10 && m_j_knot < 1.E-10
        DNurbsbasis(i,p+1+k)=0;
     elseif k_i_knot < 1.E-10
        DNurbsbasis(i,p+1+k)=-ip*DNurbsbasis(i+1,npi)/m_j_knot;
     elseif m_j_knot < 1.E-10
        DNurbsbasis(i,p+1+k)=+ip*DNurbsbasis(i,npi)/k_i_knot;
     else
        DNurbsbasis(i,p+1+k)=+ip*DNurbsbasis(i,npi)/k_i_knot...
                         -ip*DNurbsbasis(i+1,npi)/m_j_knot;
```

```
      end
    end
end
%
% Constructing NURBS p-th basis functions
for j=1:(nb-p)
     DNurbsbasis(j,p+5)=DNurbsbasis(j,p+1)*wt(j);
     ydenom=0;
     for i=1:(nb-p)
       ydenom=ydenom+DNurbsbasis(i,p+1)*wt(i);
     end
     DNurbsbasis(j,p+5)=DNurbsbasis(j,p+5)/ydenom;
end
% Constructing k-th Derivative NURBS basis functions
  SumNw=zeros(p+1);
  for i=1:p+1                                   % i=1 non-derivation
   for j=1:nb-p
     SumNw(i)=SumNw(i)+DNurbsbasis(j,p+i)*wt(j);   % Sigma(N(t)xw
(t)) term
   end
  end
  %
  for i=1:nb-p
   for k=1:3
    SumRhs=0;
    for m=1:k
        Binomial=factorial(k)/factorial(m)/factorial(k-m);
        SumRhs=Binomial*SumNw(1+k)*DNurbsbasis(i,p+4+m);
    end
    DNurbsbasis(i,p+5+k)=(DNurbsbasis(i,p+1+k)*wt(i)-SumRhs)/
    SumNw(1);
 end
 end
return
```

References

Bajaj C et al. (1995) Modeling with cubic A-patches. ACM Trans on Graphics 14:103–133

Bernstein SN (1912) Démonstration du théorèm de Weirstrass fondée sûr le calcul dès probabilités. Commun Soc Math Khrakow 12(2):1–2

Bézier PE (1986) The mathematical basis of the UNISURF CAD system. Butterworth, London

De Boor C (1987) Cutting corners always works. Comput Aided Geom Des 4(1–2):125–131

Chang G, Wu J (1981) Mathematical foundations of Bézier's technique. Comput Aided Des 13 (3):133–136

Chou JJ (1995) Higher order Bézier circles. Comput Aided Des 27(4):303–309
Gordon WJ, Riesenfeld RF (1974) Berstein-Bézier methods for the computer-aided design of free-form curves and surfaces. J Assoc Comp Mach 21(2):293–310
Lei WT, Wang SB (2009) Robust real-time NURBS path interpolators. Int J Mach Tools Manuf 49:625–633
Piegl LA, Tiller W (2003) Circle approximation using integral B-splines. Comput Aided Des 35:601–607

Further Reading

Beach RC (1991) An introduction to the curves and surfaces of computer-aided design. Van Nostrand Reinhold, New York
Bézier PE (1972) Numerical control: mathematics and applications. Wiley, New York
De Boor C (1978) A Practical Guide to Splines. Springer, New York
Farin GE (1993) Curves and surfaces for computer aided geometric design – a practical guide, 3rd edn. Academic Press, Boston
Faux ID, Pratt MJ (1981) Computational geometry for design and manufacture. Ellis Horwood Ltd., Chichester
Hoffmann CM (1989) Geometric & solid modeling. Morgan Kaufmann, San Mateo
Hoschek J, Lasser D (1993) Fundamentals of computer aided geometric design. AK Peters Ltd., Wellesley
Lorentz GG (1986) Bernstein polynomials. Chelsea Publishing Co., New York
Mortenson ME (1985) Geometric modeling. Wiley, New York
Piegl L, Tiller W (1997) The NURBS book. Springer, Berlin
Rogers DF, Adams JA (1990) Mathematical elements for computer graphics, 2nd edn. McGraw-Hill, New York
Yamaguchi F (1988) Curves and surfaces in computer aided geometric design. Springer, New York

Chapter 2
Numerical Integration

2.1 Introduction

An important aspect in the isogeometric approach, the vector and matrix formulations are requiring numerical integration scheme because the high-order polynomial presentations of the NURBS functions. The NURBS functions will be adopted as the shape functions in the formulations of a beam element which are hardly can be represented by using formulas explicitly. Our discussion of the numerical integration and their related methods is only related to the beam element formulation; for other methods, the reader should consult other references (Gradshteyn and Ryzhik 2000; Hildebrand 1956; Kaplan 1984; Ralston and Rabinowitz 1978).

2.2 Gauss-Legendre Quadrature

The Gauss-Legendre numerical integration can achieve the exact of fitness of $2n - 1$ or less by using the weighted sum of function values at specified points within the domain of integration. It approximates the following definite integration:

$$\int_{-1}^{+1} f(\xi)\ d\xi \approx \sum_{i=1}^{n} f(\xi_i)\ w_i \tag{2.1}$$

where n is the order of degree of polynomial inside the integral, ξ_i is the coordinate of Gauss point i, and w_i is the weight of Gauss point i.

The Legendre polynomials can be defined by a recursive relation as follows:

B.S. Gan, *An Isogeometric Approach to Beam Structures*,
DOI 10.1007/978-3-319-56493-7_2

$$P_n(\xi) = \frac{1}{n}\{(2n-1)\ \xi\ P_{n-1}(\xi) - (n-1)\ P_{n-2}(\xi)\} \qquad n \geq 2$$
$$\text{where} \quad P_0(\xi) = 1, \quad P_1(\xi) = \xi \tag{2.2}$$

The points ξ_i, $(i = 1,...,n)$ are the roots of (2.2) inside the range of $[-1\ 1]$ by equating the polynomial to zero as follows:

$$P_n(\xi) = 0 \qquad \text{for } n \geq 1 \tag{2.3}$$

The weights w_i are calculated from corresponding ξ_i points as

$$w_i = \frac{2(1-\xi_i^2)}{n^2\ P_{n-1}^2(\xi_i)} \qquad \text{for } n \geq 1 \tag{2.4}$$

Table 2.1 lists the points and weights for order-n Legendre polynomial up to $n = 5$.

As an example, the calculation of areas under a given function: $f(\xi) = \xi^8 + \xi^7 + \xi^6 + \xi^5 + 10$ in the interval of $[-1\ 1]$, by using the Gauss-Legendre rule, is tabulated in Table 2.2. We can see that the maximum order of the function is of the order eight. Hence, the exactness of solution can be achieved by using $2n - 1 \geq 8$; $n \geq 4.5$ (since n must be an integer, we use $n = 5$) to obtain the exact solution.

2.3 Length, Jacobian Operator, and Radius of Curvature of a Curve

What if we have to integrate an equation with real coordinates $\mathbf{x}$, different with the $\boldsymbol{\xi}$ $[-1 \quad +1]$ interval? Apparently, we will find them in most of the cases we deal with the real world. Hence, we need an operator that can scale between both coordinates systems $(\mathbf{x}, \boldsymbol{\xi})$ the so-called coordinate transformation.

In fact, the numerical integration that we should do in the finite element of a beam is a line integration scheme, not integrating the area under the line itself. The beam element components such as stiffness matrix, mass matrix, and loading vector are formed by integrating the material properties, cross section, or distributed mass along the path of a line or a curve.

In calculus, a scaling factor that relates the incremental variable of both coordinates is called the *Jacobian* operator. We denote it by J. Invoking the chain rule,

$$\frac{d}{d\boldsymbol{\xi}} = \frac{d\mathbf{x}}{d\boldsymbol{\xi}}\frac{d}{d\mathbf{x}} = J\frac{d}{d\mathbf{x}} \quad \text{where} \quad \frac{d\mathbf{x}}{d\boldsymbol{\xi}} = J \quad \text{or} \quad \frac{d\boldsymbol{\xi}}{d\mathbf{x}} = \frac{1}{J} \tag{2.5}$$

which requires that J has a non-zero determinant value. The integration becomes

Table 2.1 Tabulated list of the order-*n* Legendre polynomial points and weights up to $n = 5$

n	Point, ξ_i / Weight, w_i
1	0
	2
2	$-\sqrt{\frac{1}{3}},+\sqrt{\frac{1}{3}}$
	$1, 1$
3	$-\sqrt{\frac{3}{5}}, 0,+\sqrt{\frac{3}{5}}$
	$\frac{5}{9}, \frac{8}{9}, \frac{5}{9}$
4	$-\sqrt{\frac{3}{7}+\frac{2}{7}\sqrt{\frac{6}{5}}},-\sqrt{\frac{3}{7}-\frac{2}{7}\sqrt{\frac{6}{5}}},+\sqrt{\frac{3}{7}-\frac{2}{7}\sqrt{\frac{6}{5}}},+\sqrt{\frac{3}{7}+\frac{2}{7}\sqrt{\frac{6}{5}}}$
	$\frac{1}{2}+\frac{1}{6}\sqrt{\frac{5}{6}}, \frac{1}{2}-\frac{1}{6}\sqrt{\frac{5}{6}}, \frac{1}{2}-\frac{1}{6}\sqrt{\frac{5}{6}}, \frac{1}{2}+\frac{1}{6}\sqrt{\frac{5}{6}}$
5	$-\frac{1}{3}\sqrt{5+2\sqrt{\frac{10}{7}}},-\frac{1}{3}\sqrt{5-2\sqrt{\frac{10}{7}}},+\frac{1}{3}\sqrt{5-2\sqrt{\frac{10}{7}}},+\frac{1}{3}\sqrt{5+2\sqrt{\frac{10}{7}}}$
	$\frac{322+13\sqrt{70}}{900}, \frac{322-13\sqrt{70}}{900}, \frac{322-13\sqrt{70}}{900}, \frac{322+13\sqrt{70}}{900}$

Table 2.2 Calculation of area by using different order of Legendre polynomial

n	$area = \sum_{i=1}^{n} f(\xi_i)\ w_i$	Exact to 16 decimal places
1	20.00000000000000	20.50793650793650
2	20.09876543209880	20.50793650793650
3	20.38400000000000	20.50793650793650
4	20.49632653061220	20.50793650793650
5	20.50793650793650	20.50793650793650

$$\int_{x_1}^{x_2} f(x)\ dx = \int_{-1}^{+1} f(\xi)\quad J\ d\xi \approx \sum_{i=1}^{n} f(\xi_i)\ J\ w_i \tag{2.6}$$

For one-dimensional integration of a straight line, we can easily relate the $x = \left(\frac{1-\xi}{2}\right)x_1 + \left(\frac{1+\xi}{2}\right)x_2$ relationship. Here $[x_1\ x_2]$ is the interval of the integration along the x coordinate. We can obtain the Jacobian operator from (2.5) as

$$J = \frac{dx}{d\xi} = -\frac{1}{2}x_1 + \frac{1}{2}x_2 = \frac{x_2 - x_1}{2} \tag{2.7}$$

Where the $[x_1\ x_2]$ interval shows the length of the beam. The length and total mass of the beam can be calculated from the following equation.

$$\begin{aligned} \text{Length} &: \int_{x_1}^{x_2} dx = \int_{-1}^{+1} J\ d\xi \approx \sum_{i=1}^{n} \left(\frac{x_2 - x_1}{2}\right)\ w_i \\ \text{Mass} &: \int_{x_1}^{x_2} \rho(x)\ dx = \int_{-1}^{+1} \rho(\xi)\ J\ d\xi \approx \sum_{i=1}^{n} \rho(\xi)\left(\frac{x_2 - x_1}{2}\right)\ w_i \end{aligned} \tag{2.8}$$

In the finite element context, the length of a curve segment, Jacobian operator, and curvature, respectively, can be given as

$$
\begin{aligned}
C(\xi) &= \int_{\xi_0=-1}^{\xi_0=\xi} \sqrt{\left(\frac{dx}{d\xi_0}\right)^2 + \left(\frac{dy}{d\xi_0}\right)^2}\, d\xi_0 \\
J &= \frac{dC}{d\xi} = \sqrt{\left(\frac{dx}{d\xi}\right)^2 + \left(\frac{dy}{d\xi}\right)^2} \\
\kappa(\xi) &= \frac{\left|\frac{dx}{d\xi}\frac{d^2y}{d\xi^2} - \frac{dy}{d\xi}\frac{d^2x}{d\xi^2}\right|}{J^3}
\end{aligned}
\tag{2.9}
$$

In case of a circular curved line, we can define the x-y coordinates by the trigonometric functions $x = R\ Cos(\varphi), y = R\ Sin(\varphi)$ and apply the chain rule from the relationship $\varphi = \frac{\varphi_1+\varphi_2}{2} + \frac{\varphi_1-\varphi_2}{2}\xi$, we can obtain the Jacobian operator from (2.9) as follows:

$$
\begin{aligned}
&\frac{dx}{d\xi} = -\left(\frac{\varphi_1 - \varphi_2}{2}\right) R\ Sin(\varphi)\ ;\ \frac{dy}{d\xi} = \left(\frac{\varphi_1 - \varphi_2}{2}\right) R\ Cos(\varphi) \\
&J = \left(\frac{\varphi_1 - \varphi_2}{2}\right) R
\end{aligned}
\tag{2.10}
$$

Where the $[\varphi_1\ \ \varphi_2]$ interval shows the arc segment length of the beam. The length and total mass of the beam can be calculated from the following equation.

$$
\begin{aligned}
\text{Length} &: \int_{x_1}^{x_2} dx = \int_{-1}^{+1} J\ d\xi \approx \sum_{i=1}^{n} \left(\frac{\varphi_1 - \varphi_2}{2}\right) R\ w_i \\
\text{Mass} &: \int_{x_1}^{x_2} \rho(x)\ dx = \int_{-1}^{+1} \rho(\xi)\ J\ d\xi \approx \sum_{i=1}^{n} \rho(\xi) \left(\frac{\varphi_1 - \varphi_2}{2}\right) R\ w_i
\end{aligned}
\tag{2.11}
$$

For a circle, where the radius of curvature is constant along the path, we have $\kappa(\xi) = {}^1/_R$. In fact, (2.9) can generally be used for all kind of curves, including the straight line by setting a zero value of curvature.

For a third-order polynomial curved line $y = x^3 - 1$, we can approximate the x coordinate related by a linear function $x = \left(\frac{1-\xi}{2}\right)x_1 + \left(\frac{1+\xi}{2}\right)x_2$ and by applying the chain rule for the y coordinate,

$$
\frac{dy}{dx} = 3x^2\ ;\ \frac{dy}{d\xi} = \frac{dy}{dx} \times \frac{dx}{d\xi} = 3x^2 \left(\frac{x_2 - x_1}{2}\right)
\tag{2.12}
$$

where the $[x_1\ \ x_2]$ interval shows the arc segment length of the beam. The Jacobian operator can be obtained from (2.9).

The length and total mass of the beam can be calculated from the following equation.

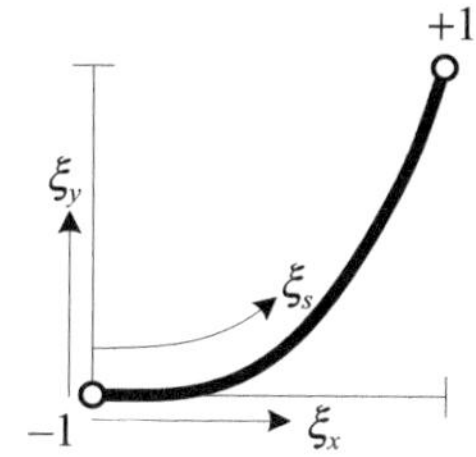

Fig. 2.1 Several possible coordinate transformations along the path of the curve in the Gauss-Legendre quadrature

$$\begin{array}{lll} \text{Length} & : & \int_{x_1}^{x_2} dx = \int_{-1}^{+1} J \ d\xi \approx \sum_{i=1}^{n} J(\xi)R \ w_i \\ \text{Mass} & : & \int_{x_1}^{x_2} \rho(x) \ dx = \int_{-1}^{+1} \rho(\xi) \ J(\xi) \ d\xi \approx \sum_{i=1}^{n} \rho(\xi)J(\xi)w_i \end{array} \tag{2.13}$$

In the above equation, the approximation of the x-abscissa to have a linear relationship with ξ is not accurate. To get an accurate result, the ξ parameter must be measured along the path of the curve, which can be done by doing the numerical integration scheme for computing the length of the curve in advance. As shown in Fig. 2.1, we can select either the x, y, or s axes optionally to be represented by the local coordinate ξ using the Gauss-Legendre quadrature integration. In a general curve, most probably the scaling factors among the axes are different. Hence, we get inaccurate results. The most convenient situation is when we have an explicit formula for the $s(x,y)$ as a function of x and y coordinates.

MATLAB code Gaussquadrature.m (Sect. 2.3.1) uses the function Legendre.m (Sect. 2.3.2), to integrate numerically the length of curves (Fig. 2.2) and their total mass from distributed density function which is defined by a polynomial equation by using (2.6) and (2.9). The computed lengths and total mass are given in Table 2.3. In the program, the calculation of length can be done by letting the variable ρ (ξ) is equal to one.

It can be noted that from the beginning of this book, the interval of the local coordinate ξ is kept unchanged between the range of [−1 + 1] to have a smooth transition between equations and formulations across the chapters.

2.3.1 Gaussquadrature Program List

```
% Gaussquadrature.m
% Length and Mass calculation of curves
% Given density of mass per unit length
clear all;clc;
ng=3;
Gpw = Legendre(ng);
%
% Distributed material on a straight line
x=[-2 1];
```

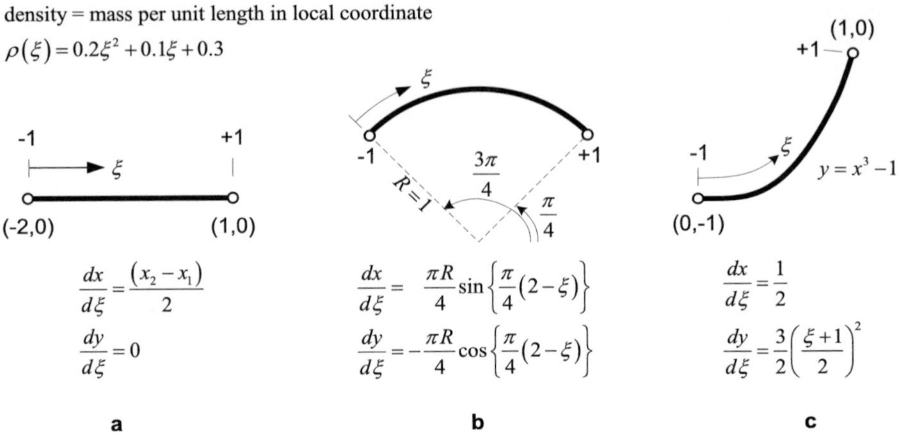

Fig. 2.2 Numerical integration of several curves and their total mass calculation: (**a**) straight line; (**b**) arc of a circle; (**c**) third-order polynomial curve

Table 2.3 Numerical integration of length of the curves and their total mass

Type of curve	Length	Total mass	Gauss point used
(a) Straight line	3.0000	1.1000	2
(b) Arc of a circle	1.5708	0.5760	2
(c) Third-order polynomial	1.5478	0.6162	5

```
dxdxi=(x(2)-x(1))/2; dydxi=0;
Jac=sqrt(dxdxi^2+dydxi^2);
C1=0;
for i=1:ng
 rho=0.2*Gpw(i,1)^2+0.1*Gpw(i,1)+0.3;
 %rho=1; % to calculate the length
 C1=C1+rho*Gpw(i,2)*Jac; % rho(xi) along the straight line
end
%
% Distributed material on a 90 deg arc of a circle
r=1;
Jac=pi*r/4;
C2=0;
for i=1:ng
 rho=0.2*Gpw(i,1)^2+0.1*Gpw(i,1)+0.3;
 %rho=1; % to calculate the length
 C2=C2+rho*Gpw(i,2)*Jac; % rho(xi) along the arc of a circle
end
%
% Distributed material on a 3rd order polynomial curve
C3=0;
```

```
for i=1:ng
 Jac=sqrt((1/2)^2+9/4*((Gpw(i,1)+1)/2)^4);
 rho=0.2*Gpw(i,1)^2+0.1*Gpw(i,1)+0.3;
 %rho=1; % to calculate the length
 C3=C3+rho*Gpw(i,2)*Jac; % rho(xi) along the 3rd order polynomial
curve
end
% p = order
```

2.3.2 Legendre Function List

```
function Gpw = Legendre(ng)
% ng = number of Gauss points
% Gauss point is stored in Gpw(n,1)
% Gauss weight is stored in Gpw(n,2)
%--------------------------------
Gpw=zeros(ng,2);
switch ng
 case 1
   Gpw(1,1)=0;
   Gpw(1,2)=2;
 case 2
   Gpw(1,1)=-sqrt(1/3);
   Gpw(1,2)=1;
   Gpw(2,1)=+sqrt(1/3);
   Gpw(2,2)=1;
 case 3
   Gpw(1,1)=-sqrt(3/5);
   Gpw(1,2)=5/9;
   Gpw(2,1)=0;
   Gpw(2,2)=8/9;
   Gpw(3,1)=+sqrt(3/5);
   Gpw(3,2)=5/9;
 case 4
   Gpw(1,1)=-sqrt(3/7+2/7*sqrt(6/5));
   Gpw(1,2)=(18-sqrt(30))/36;
   Gpw(2,1)=-sqrt(3/7-2/7*sqrt(6/5));
   Gpw(2,2)=(18+sqrt(30))/36;
   Gpw(3,1)=+sqrt(3/7-2/7*sqrt(6/5));
   Gpw(3,2)=(18+sqrt(30))/36;
   Gpw(4,1)=+sqrt(3/7+2/7*sqrt(6/5));
   Gpw(4,2)=(18-sqrt(30))/36;
```

```
case 5
  Gpw(1,1)=-1/3*sqrt(5+2*sqrt(10/7));
  Gpw(1,2)=(322-13*sqrt(70))/900;
   Gpw(2,1)=-1/3*sqrt(5-2*sqrt(10/7));
   Gpw(2,2)=(322+13*sqrt(70))/900;
   Gpw(3,1)=0;
   Gpw(3,2)=128/225;
   Gpw(4,1)=+1/3*sqrt(5-2*sqrt(10/7));
   Gpw(4,2)=(322+13*sqrt(70))/900;
   Gpw(5,1)=+1/3*sqrt(5+2*sqrt(10/7));
   Gpw(5,2)=(322-13*sqrt(70))/900;
case 6
   Gpw(1,1) = -0.932469514203151938982;
   Gpw(2,1) = -0.661209386466264592509;
   Gpw(3,1) = -0.238619186083196932469;
   Gpw(4,1) = 0.238619186083196932469;
   Gpw(5,1) = 0.661209386466264592509;
   Gpw(6,1) = 0.932469514203151938982;
   Gpw(1,2) = 0.171324492379171867455;
   Gpw(2,2) = 0.360761573048138139704;
   Gpw(3,2) = 0.467913934572691007877;
   Gpw(4,2) = 0.467913934572691007877;
   Gpw(5,2) = 0.360761573048138139704;
   Gpw(6,2) = 0.171324492379171867455;
case 7
   Gpw(1,1) = -0.949107912342758486214;
   Gpw(2,1) = -0.741531185599394682074;
   Gpw(3,1) = -0.405845151377397184156;
   Gpw(4,1) = 0.000000000000000000000;
   Gpw(5,1) = 0.405845151377397184156;
   Gpw(6,1) = 0.741531185599394682074;
   Gpw(7,1) = 0.949107912342758486214;
   Gpw(1,2) = 0.129484966168870452732;
   Gpw(2,2) = 0.279705391489274882221;
   Gpw(3,2) = 0.381830050505118893749;
   Gpw(4,2) = 0.417959183673469387755;
   Gpw(5,2) = 0.381830050505118893749;
   Gpw(6,2) = 0.279705391489274882221;
   Gpw(7,2) = 0.129484966168870452732;
case 8
   Gpw(1,1) = -0.960289856497536509162;
   Gpw(2,1) = -0.796666477413626949956;
   Gpw(3,1) = -0.525532409916328990763;
   Gpw(4,1) = -0.183434642495649807836;
   Gpw(5,1) = 0.183434642495649807836;
```

```
    Gpw(6,1) = 0.525532409916328990763;
    Gpw(7,1) = 0.796666477413626949956;
    Gpw(8,1) = 0.960289856497536509162;
    Gpw(1,2) = 0.101228536290370022005;
    Gpw(2,2) = 0.222381034453372634246;
    Gpw(3,2) = 0.313706645877887267061;
    Gpw(4,2) = 0.362683783378361979375;
    Gpw(5,2) = 0.362683783378361979375;
    Gpw(6,2) = 0.313706645877887267061;
    Gpw(7,2) = 0.222381034453372634246;
    Gpw(8,2) = 0.101228536290370022005;
case 9
    Gpw(1,1) = -0.968160239507625086598;
    Gpw(2,1) = -0.836031107326636102552;
    Gpw(3,1) = -0.613371432700592578104;
    Gpw(4,1) = -0.324253423403808971326;
    Gpw(5,1) = 0.000000000000000000000;
    Gpw(6,1) = 0.324253423403808971326;
    Gpw(7,1) = 0.613371432700592578104;
    Gpw(8,1) = 0.836031107326636102552;
    Gpw(9,1) = 0.968160239507625086598;
    Gpw(1,2) = 0.081274388361599606396;
    Gpw(2,2) = 0.180648160694854311268;
    Gpw(3,2) = 0.260610696402924285138;
    Gpw(4,2) = 0.312347077040002744348;
    Gpw(5,2) = 0.330239355001259763165;
    Gpw(6,2) = 0.312347077040002744348;
    Gpw(7,2) = 0.260610696402924285138;
    Gpw(8,2) = 0.180648160694854311268;
    Gpw(9,2) = 0.081274388361599606396;
case 10
    Gpw(1,1) = -0.973906528517169189864;
    Gpw(2,1) = -0.865063366688986756738;
    Gpw(3,1) = -0.679409568299023991446;
    Gpw(4,1) = -0.433395394129247379933;
    Gpw(5,1) = -0.148874338981631215706;
    Gpw(6,1) = 0.148874338981631215706;
    Gpw(7,1) = 0.433395394129247379933;
    Gpw(8,1) = 0.679409568299023991446;
    Gpw(9,1) = 0.865063366688986756738;
    Gpw(10,1)= 0.973906528517169189864;
    Gpw(1,2) = 0.066671344308758311881;
    Gpw(2,2) = 0.149451349150555209279;
    Gpw(3,2) = 0.219086362515984566833;
    Gpw(4,2) = 0.269266719309995757214;
```

```
    Gpw(5,2) = 0.2955242247147528654022;
    Gpw(6,2) = 0.2955242247147528654022;
    Gpw(7,2) = 0.2692667193099957572143;
    Gpw(8,2) = 0.2190863625159845668333;
    Gpw(9,2) = 0.1494513491505552092793;
    Gpw(10,2) = 0.0666713443087583118811;
end
return
```

2.4 Sinusoidal Curve Example

MATLAB code Sinusoidalcurve.m (Sect. 2.4.1) uses the functions Nurbs.m (Sect. 1.3.10) and DNurbsLeibnitz.m (Sect. 1.3.14), to draw a sinusoidal curve (2.14), its derivative, Jacobian values, and curvature along the length of a curve.

$$y = 1.5\sin[x] \quad , \quad 0 \le x \le 2\pi \tag{2.14}$$

Rewriting the above equation into the Gaussian coordinate form,

$$y = 1.5\sin[\pi(\xi + 1)] \quad , \quad -1 \le \xi \le +1 \tag{2.15}$$

with

$x = \pi(\xi + 1)$

Jacobian, $J(\xi) = \frac{dy}{d\xi} = \frac{dy}{dx}\cdot\frac{dx}{d\xi} = 1.5\pi\cos[\pi(\xi + 1)]$.

Curvature, $\kappa(\xi) = \frac{\left|\frac{dx}{d\xi}\frac{d^2y}{d\xi^2} - \frac{dy}{d\xi}\frac{d^2x}{d\xi^2}\right|}{J(\xi)^3} = \frac{\left|-\pi^2 J(\xi)\right|}{J(\xi)^3} = \frac{\pi^2}{J(\xi)^2}$.

Figure 2.3 depicts the sinusoidal curve ($p = 8$), its derivative curve, Jacobian values, and curvature along the length of curve.

2.4.1 *Sinusoidalcurve Program List*

```
% Sinusoidalcurve.m
% Calculating the Jacobian and Radius of Curvature of a sinusoidal
  curve
% p = degree of Bspline curves
clear all;clc;
p=8;
```

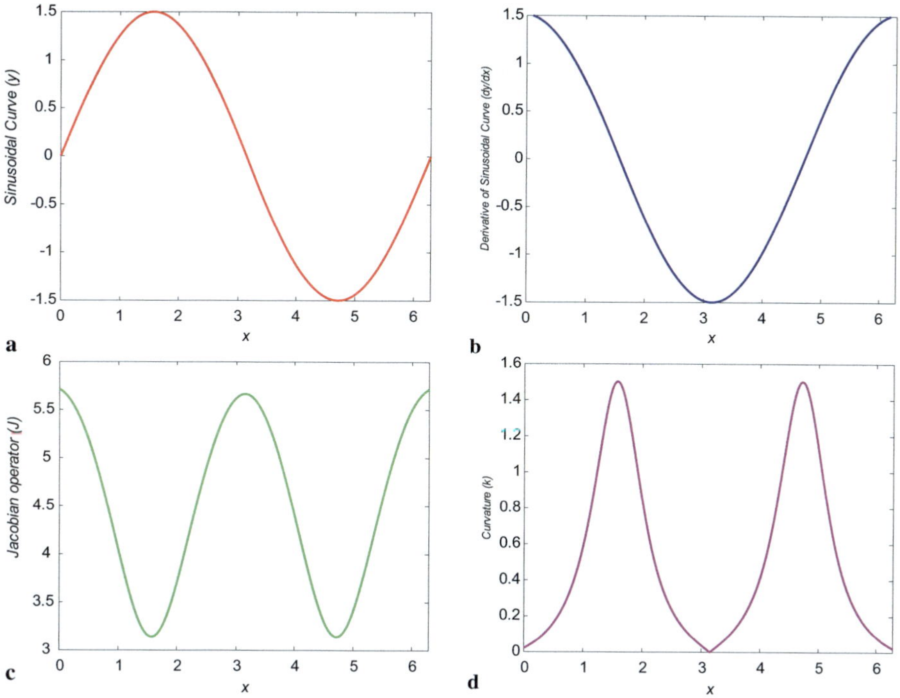

Fig. 2.3 Sinusoidal curve example: (**a**) curve; (**b**) derivative of curve; (**c**) Jacobian; (**d**) curvature

```
xi=zeros(1,p+1);
yi=zeros(1,p+1);
c1=(0.2*sqrt(2)/(sqrt(2)-1));
for ip=1:p+1
 xi(1,ip)=2*pi*(ip-1)/p;
 ksi=(xi(1,ip)-pi)/pi;
 yi(1,ip)=1.5*sin(pi*(ksi+1));
end
xyi=transpose([xi yi]);
% Knot vector
knot=[-1*ones(1,p+1) 1*ones(1,p+1)];
knot=[knot max(knot)*ones(1,p)];
nb=size(knot,2)-p-1;          % max number of basis
                              functions
% Parameter t
t=-1:2/p:1;
nt=size(t,2);
wt=ones(1,p+1);
Sb=zeros(nt,1);               % BsplineBasisPlot (order,basis)
tmi=zeros(nt,p+1);
```

```
for i=0:nb-p-1
 for dt=1:nt
   Nurbsbasis = Nurbs(p,knot,wt,t(dt));
   tmi(dt,i+1)=Nurbsbasis(i+1,p+2);
 end
end
% [tm] matrix
tm=[tmi zeros(nb-p,nb-p);zeros(nb-p,nb-p) tmi];
% Coefficients Px's and Py's calculation
pxy=tm\xyi;
px=pxy(1:nt);       % n+1 control points
py=pxy(nt+1:2*nt); % n+1 control points
%nb=size(knot,2)-p-1; % max number of basis functions
ndt=100;               % steps of parameter t
t1=min(knot); t2=max(knot);
t=t1:(t2-t1)/ndt:t2;
nt=size(t,2);
wt=ones(nb-p);          % Uniform weights
curvx=zeros(nt,1);      % Nurbs variables
curvy=zeros(nt,1); curvyd=zeros(nt,1);
xx=zeros(nt);
Jac=zeros(nt);
curva=zeros(nt,1);
% Calculating the Jacobian and Radius of Curvature
for it=1:nt
 Nurbsbasis = DNurbsLeibnitz(p,knot,wt,t(it));
 dx=0;  dy=0;
 ddy=0; ddx=0;
 for ib=1:nb-p                         % Number of basis function
 dx=dx+Nurbsbasis(ib,p+7)*px(ib);   % dx
 dy=dy+Nurbsbasis(ib,p+7)*py(ib);   % dy
 ddx=ddx+Nurbsbasis(ib,p+8)*px(ib); % dx
 ddy=ddy+Nurbsbasis(ib,p+8)*py(ib); % dy
 end
 Jac(it) =sqrt(dx^2+dy^2);
 curva(it)=abs(dx*ddy-dy*ddx)/Jac(it)^3;
end
% Drawing the Curve and its Derivative
for it=1:nt                 % ksi
 dy=0; dx=0;
 for ib=1:nb-p              % Number of basis function
 Nurbsbasis = DNurbsLeibnitz(p,knot,wt,t(it));
 curvx(it) =curvx(it) +Nurbsbasis(ib,p+5)*px(ib);
 curvy(it)=curvy(it)+Nurbsbasis(ib,p+5)*py(ib); % NonDerivative
 dx=dx+Nurbsbasis(ib,p+7)/Jac(it)*px(ib);
```

```
  dy=dy+Nurbsbasis(ib,p+7)/Jac(it)*py(ib);
  end
  curvyd(it)=dy/dx; % 1st Derivative
end
figure(1); plot(curvx,curvy,'-k','LineWidth',2);hold all;
axis([0.0 2*pi -1.5 1.5]);
xlabel('\itx'); ylabel('\itSinusoidal Curve (y)');
figure(2); plot(curvx,curvyd,'-k','LineWidth',2);hold all;
axis([0.0 2*pi -1.5 1.5]);
xlabel('\itx'); ylabel('\itDerivative of Sinusoidal Curve
(dy/dx)');
figure(3); plot(curvx,Jac,'-k','LineWidth',2);hold all;
axis([0.0 2*pi 3.0 6.0]);
xlabel('\itx'); ylabel('\itJacobian operator (J)');
figure(4); plot(curvx,curva,'-k','LineWidth',2);hold all;
axis([0.0 2*pi 0.0 1.6]);
xlabel('\itx'); ylabel('\itCurvature (k)');
```

Further Reading

Gradshteyn IS, Ryzhik IM (2000) Table of integrals, series, and products, 6th edn. Academic Press, San Diego

Hildebrand FB (1956) Introduction to numerical analysis. McGraw-Hill, New York

Kaplan W (1984) Advanced calculus, 3rd edn. Addison-Wesley, Reading

Ralston A, Rabinowitz P (1978) A first course in numerical analysis. McGraw-Hill, London

Chapter 3
Finite Element Formulation of Beam Elements

3.1 What Is a Finite Element?

To illustrate the concept of the finite element method (FEM), let us go back to the era of 250 BC, when the famous ancient Greek mathematician, physicist, engineer, inventor, and astronomer Archimedes studied rectification of a circle. The problem is "to find a relationship between the perimeter P and the diameter D of a circle." The answer to this problem is to find a numerical value for π from the following equation:

$$\begin{aligned} &P = \pi\ D \\ &\text{where, } \pi = \textstyle\sum_{i=1}^{n} \sin(180/n) = n \ \ \sin(180/n) \end{aligned} \tag{3.1}$$

The perimeter P of a circle with diameter D as shown in Fig. 3.1a will be approximated by n polygons with equal side lengths (Fig. 3.1b) that can be calculated from a simple trigonometry relationship (Fig. 3.1c).

Values of π calculated from (3.1) are shown in Table 3.1. Since the perimeter P of the circle can be approximated by summing the length of n-polygon sides, we need to increase the value of n more than one million divisions to reach the exact value of π to 12 decimal places. This example has been presented in most of the FEM textbooks and literature to teach the finite element concept (Bathe 1982; Cook 1981; Fung 1969; Shames and Dym 1985; Washizu 1982).

3.2 Beam Element in the Real World

Most of the structural components in man-made engineering structures are composed of beams. A beam is a long-block-like structural member whose primarily function is to support transverse loading and distribute it to the supports (Fig. 3.2).

B.S. Gan, *An Isogeometric Approach to Beam Structures*,
DOI 10.1007/978-3-319-56493-7_3

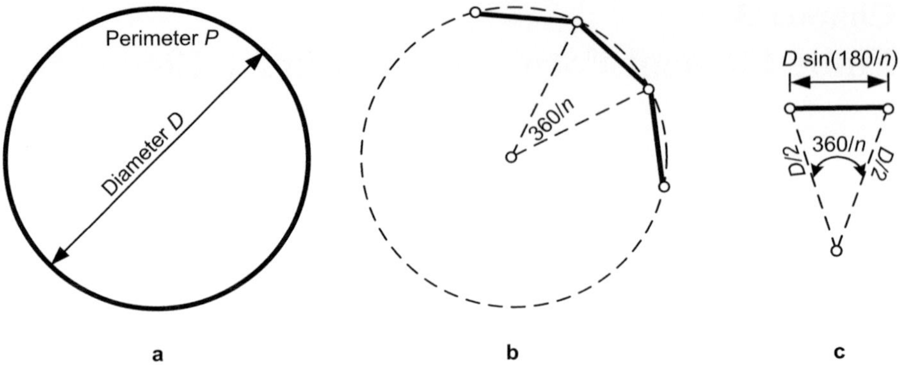

Fig. 3.1 The find "π" problem: (**a**) circle; (**b**) polygon equal n sides; (**c**) trigonometric relationship

Table 3.1 Rectification of perimeter of a circle by n-polygon length ("Archimedes FEM")

n	$\pi = n \sin(180/n)$	Exact to 12 decimal places
1	0.00000000000	3.14159265359
10	3.09016994375	3.14159265359
100	3.14107590781	3.14159265359
1000	3.14158748588	3.14159265359
10,000	3.14159260191	3.14159265359
1,000,000	3.14159265358	3.14159265359
1,037,803	3.14159265359	3.14159265359

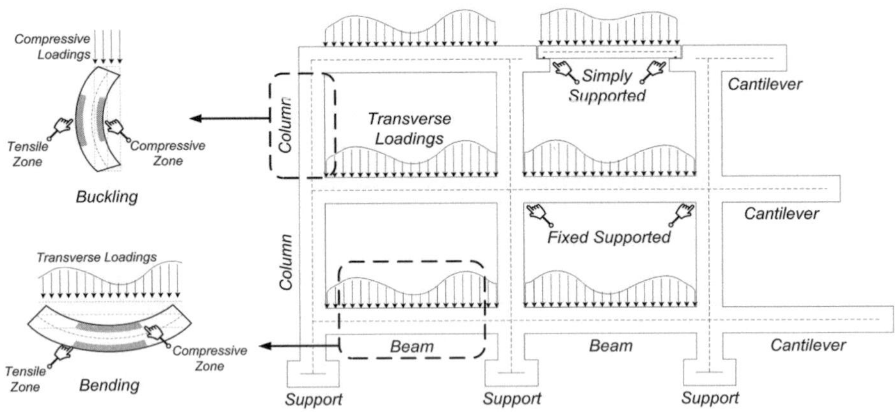

Fig. 3.2 Schematic modeling a beam in a building structure

The beam resists the transverse loading mainly through a bending mechanism. A bending mechanism produces a compressive zone on one side, and a tensile zone on the other side of a beam depends on the loading configurations. In a building structure, a column is also designed based on the classical beam theory with an exception that it can undergo a buckling phenomenon if loaded by an excessive compressive loading. A beam can resist a combination of loading actions. When a beam is supported at both of its ends where they are permitted to rotate, we call the beam as simply supported. If the beam is fixed at one end and freed at the other end, we call it a cantilever beam.

3.3 Beam in Finite Element Context

In the finite element context, a beam is modeled by a line which represents the axis of the beam in the longitudinal direction or perpendicular to its cross section (Fig. 3.3). The line of a beam is called an element. A typical beam element has two nodal points or nodes at both extremities of the line. Each node is given freedom to move or rotate; hence, we call it a *degree of freedom* (DOF). Sometimes, the degree in the DOF can be misinterpreted by order of equations. It is not true. The term degree is only showing some freedoms to displace or to rotate at the node. The type of freedom at a node can either be a displacement or rotation or a combination between both of them. It is a standard method of the numerical computation; the formulations of beam element in finite element analysis are presented in the matrices and vectors notations. The simplest and common planar beam element has two nodes at both ends, and each node has a horizontal (tangential) displacement, a vertical (transversal) displacement, and a counterclockwise (CCW) rotation as the degree of freedom. Hence, there are three DOFs at each node. Thus, six DOFs are attributed to a beam element. In vector notation, these six DOFs are called as a *generalized displacement* which is given as follows:

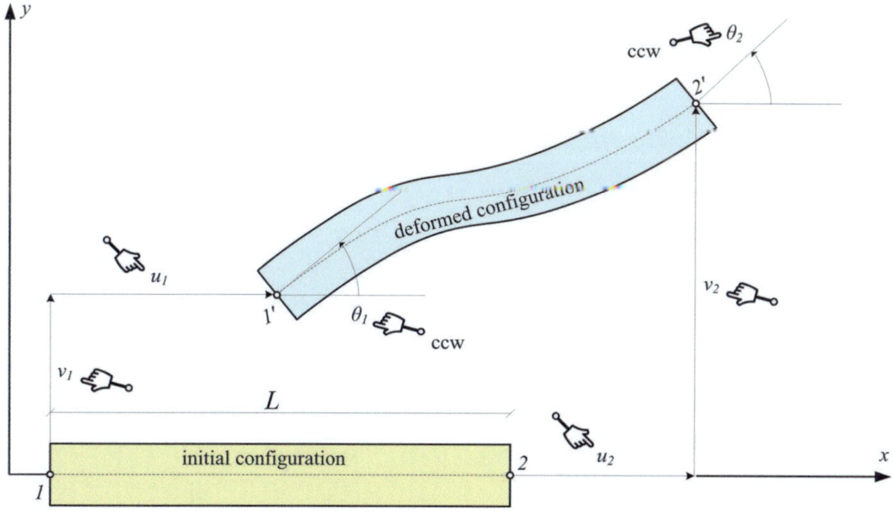

Fig. 3.3 Beam element in finite element context

$$\mathbf{d} = [u_1 \ v_1 \ \theta_1 \ u_2 \ v_2 \ \theta_2]^T \tag{3.2}$$

It will be discussed later in the following sections that the standard beam element formulation is based on the third-order polynomial assumption. When the initial shape of a beam is not straight or in vibration or has dynamic analyses problems, higher-order polynomial must be assumed. Hence, to increasing the accuracy of the calculation, the beam element must be divided into smaller divisions.

3.4 Theory of Beams

The objective of this section is to refresh the theory behind the formulation of beams. The theory of beams is mainly divided whether the effects of shear deformation are included in the calculation or neglected because the shear effect is negligible. In the following sections, we discuss first the fundamental theories of beam and matrix formulations of those beams.

3.4.1 *Straight Euler-Bernoulli Beam*

The underlying assumption in Euler-Bernoulli beam theory is that the plane of the cross section remains perpendicular to the beam axis during the deformation. This implies that the rotation of the cross section of the beam is equal to the slope of the beam axis. As illustrated in Fig. 3.4, this kinematic assumption of a beam leads to

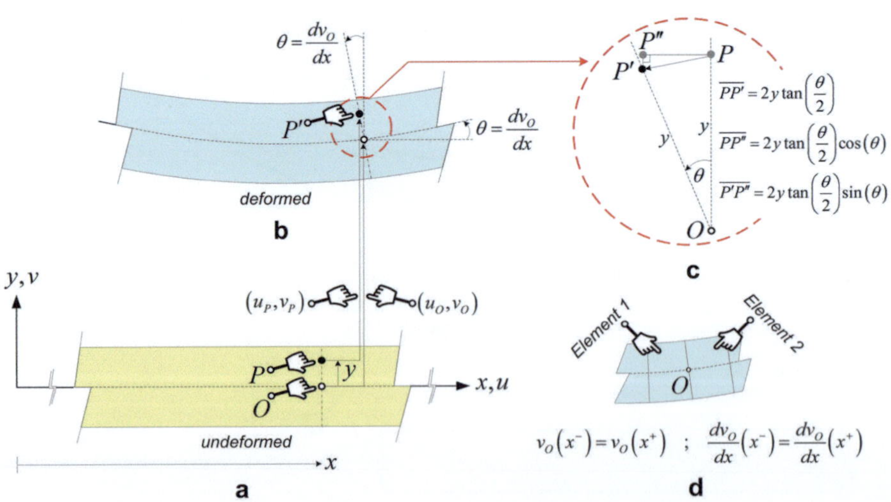

Fig. 3.4 Conceptual kinematic of a straight Euler-Bernoulli beam element

the well-known beam bending governing differential equation in which the transverse displacement v is the only variable. Therefore, the continuity conditions between two connected elements at point O require that v and $\frac{dv}{dx}$ are continuous (Fig. 3.4d). Point O is resting on the axis of the beam.

Under the small displacement assumption, we can apply the following approximations: $\sin(\theta) \approx \tan(\theta) \approx \theta$; and $\cos(\theta) \approx 1$; $\theta^2 \approx 0$.

We can observe from Fig. 3.4c that during the deformation of the beam, an arbitrary point P is displaced and rotated by the following assumed displacement fields:

$$
\begin{aligned}
u_P &= u_O - 2y\tan\left(\frac{\theta}{2}\right)\cos(\theta) \approx u_O - 2y\frac{\theta}{2}1 \approx u_O - y\theta \\
v_P &= v_O + y - y - 2y\tan\left(\frac{\theta}{2}\right)\sin(\theta) \approx v_O - 2y\frac{\theta}{2}\theta \approx v_O
\end{aligned}
\tag{3.3}
$$

where u_o and u_P are the horizontal displacements of points O and P, v_o and v_P are the transversal displacements of points O and P, x and y are the coordinates of point P before the deformation, and θ is the rotation of the cross section of the beam which is equal to slope the beam at point O.

3.4.2 *Straight Timoshenko Beam*

In the Timoshenko beam theory, the plane of the cross section is considered to not remain perpendicular to the beam axis during the deformation. Therefore, the rotation of the cross section of the beam is not equal to the slope of the beam axis. As illustrated in Fig. 3.5, this kinematic assumption of a beam leads to the well-known Timoshenko beam governing differential equation in which the shear deformation effects are taken into account. This shear deformation is considered as a *shearing relieving action* of the deformed cross section of the beam (Fig. 3.5b). As we can see from the figure, the physical explanation of shear deformation can be considered as the reaction of the cross section which is trying to rotate back from the beam axis perpendicular line after being bent.

The continuity conditions between two connected elements at point O require that v and θ be continuous (Fig. 3.5d).

We can observe from Fig. 3.5c, during the deformation of the beam, an arbitrary point P is displaced and rotated by the following assumed displacement fields:

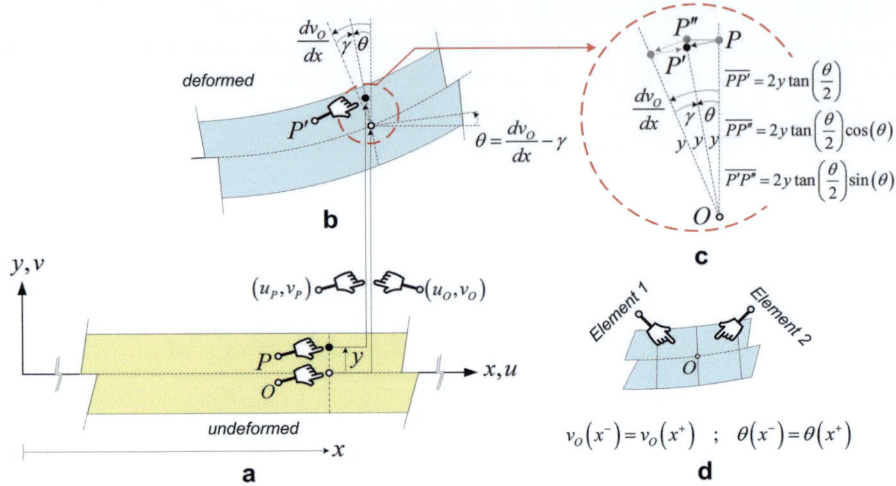

Fig. 3.5 Conceptual kinematic of a straight Timoshenko beam element

$$
\begin{aligned}
u_P &= u_O - 2y\tan\left(\frac{\theta}{2}\right)\cos(\theta) \approx u_O - 2y\frac{\theta}{2}1 \approx u_O - y\theta \\
v_P &= v_O + y - y - 2y\tan\left(\frac{\theta}{2}\right)\sin(\theta) \approx v_O - 2y\frac{\theta}{2}\theta \approx v_O \\
\theta &= \frac{dv_O}{dx} - \gamma
\end{aligned}
\tag{3.4}
$$

where u_o and u_P are the horizontal displacements of points O and P, v_o and v_P are the transversal displacements of points O and P, x and y are the coordinates of point P before the deformation, θ is the rotation of the cross section of the beam which is not equal to slope the beam at point O, and γ is the shear deformation.

3.4.3 General Curved Euler-Bernoulli Beam

The kinematic assumption of a general curved Euler-Bernoulli beam will be discussed in this section. From there, we can use the formulas to analyze the circular, elliptic, sinusoidal, parabolic, and other geometric representations of beams on the plane. The circular curved beam is a particular case of the general curved beam where the radius of curvature is constant along the path of the beam element.

As illustrated in Fig. 3.6, this kinematic assumption of a beam leads to the well-known beam bending governing differential equation in which the transversal displacement v is the only variable. Therefore, the continuity conditions between two connected elements at node O require that v and $\frac{dv}{dx}$ are continuous (Fig. 3.6d).

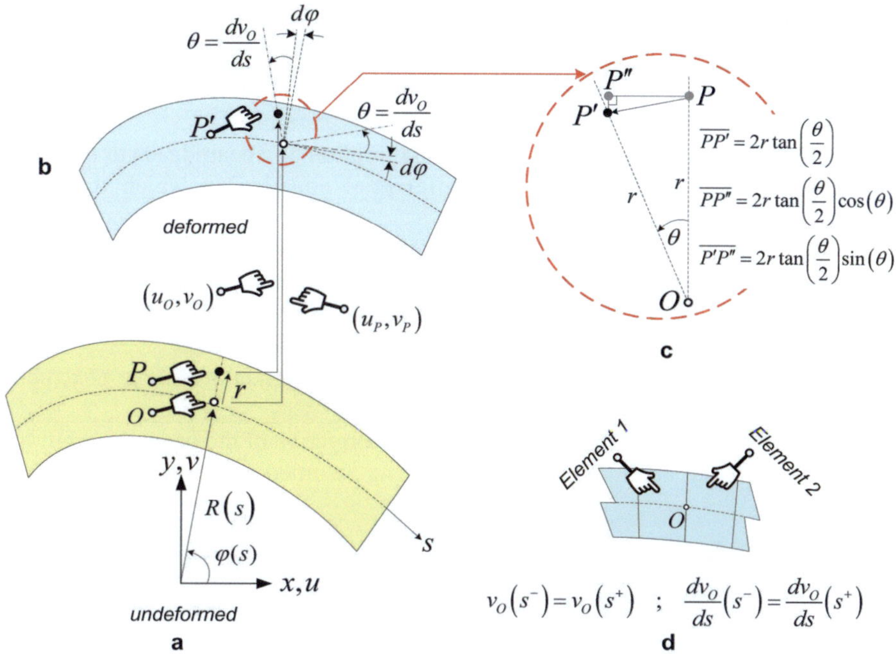

Fig. 3.6 Conceptual kinematic of a general curved Euler-Bernoulli beam element

Under the small displacement assumption, we can apply the following approximations: $\sin(\theta) \approx \tan(\theta) \approx \theta$; and $\cos(\theta) \approx 1$; $\theta^2 \approx 0$.

We can observe from Fig. 3.6c that during the deformation of the beam, an arbitrary point P is displaced and rotated by the following assumed displacement fields:

$$\begin{aligned} u_P &= u_O - 2r\tan\left(\frac{\theta}{2}\right)\cos(\theta) - 2v_O\tan\left(\frac{d\varphi}{2}\right)\cos(d\varphi) \\ &\approx u_O - 2r\frac{\theta}{2}1 - 2v_O\frac{d\varphi}{2} \approx u_O - r\theta - v_O d\varphi \\ v_P &= v_O + r - r - 2r\tan\left(\frac{\theta}{2}\right)\sin(\theta) - 2u_O\tan\left(\frac{d\varphi}{2}\right)\cos(d\varphi) \\ &\approx v_O - 2r\frac{\theta}{2}\theta - 2u_O\frac{d\varphi}{2} \approx v_O - u_O d\varphi \end{aligned} \tag{3.5}$$

where u_o and u_P are the tangential displacements of the point O and P; v_o and v_P are the transversal deflections of the point O and P; r and s are the radial and tangential coordinates of point P before the deformation; θ is the rotation of the cross section of the beam which is equal to slope the beam at point O; φ is the position angle of point O.

3.4.4 General Curved Timoshenko Beam

Following the same kinematic assumption with the straight beam element, a general curved Timoshenko beam will be discussed in this section.

As illustrated in Fig. 3.7, this kinematic assumption of a beam leads to the well-known Timoshenko beam governing differential equation in which the shear deformation effects are taken into account.

Under the small displacement assumption, we can apply the following approximations: $\sin(\theta) \approx \tan(\theta) \approx \theta$; and $\cos(\theta) \approx 1$; $\theta^2 \approx 0$.

We can observe from Fig. 3.7c during the deformation of the beam, an arbitrary point P is displaced by the following assumed displacement fields.

The continuity conditions between two connected elements at node O require that v and θ are continuous (Fig. 3.7d).

We can observe from Fig. 3.7c, during the deformation of the beam, an arbitrary point P is displaced and rotated by the following assumed displacement fields:

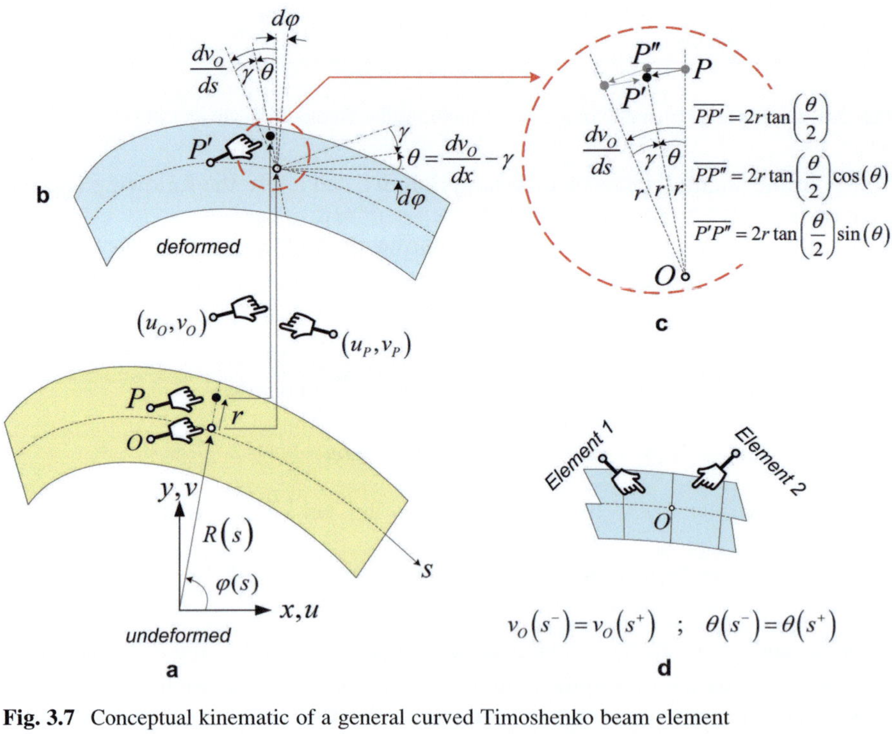

Fig. 3.7 Conceptual kinematic of a general curved Timoshenko beam element

$$\begin{aligned} u_P &= u_O - 2r\tan\left(\frac{\theta}{2}\right)\cos(\theta) - 2v_O\tan\left(\frac{d\varphi}{2}\right)\cos(d\varphi) \\ &\approx u_O - 2r\frac{\theta}{2}1 - 2v_O\frac{d\varphi}{2} \approx u_O - r\theta - v_O d\varphi \\ v_P &= v_O + r - r - 2r\tan\left(\frac{\theta}{2}\right)\sin(\theta) - 2u_O\tan\left(\frac{d\varphi}{2}\right)\cos(d\varphi) \\ &\approx v_O - 2r\frac{\theta}{2}\theta - 2u_O\frac{d\varphi}{2} \approx v_O - u_O d\varphi \end{aligned} \tag{3.6}$$

where u_o and u_P are the tangential displacements of the point O and P; v_o and v_P are the transversal deflections of the point O and P; r and s are the radial and tangential coordinates of point P before the deformation; θ is the rotation of the cross section of the beam which is equal to slope the beam at point O; φ is the position angle of point O; γ is the shear deformation.

3.5 Governing Equations of Beam Elements

In this section, the governing equations of the beam element are derived based on Hamilton's principle. Hamilton's principle is a generalization of the virtual work energy principle for obtaining the equilibrium equations in the dynamic system. Hence, the governing equations of beam element formulations can be obtained for static and dynamic systems.

3.5.1 Straight Euler-Bernoulli Beam

To generalize the displacements and rotation at point P, the subindexing of the point O in (3.3) as the reference of deformation will be eliminated. The assumed displacement of the Euler-Bernoulli beam can be rewritten as

$$\begin{aligned} u_P &= u - y\theta \\ v_P &= v \end{aligned} \tag{3.7}$$

From the theory of elasticity, we can obtain the strain component without shear deformation as

$$\begin{aligned}\varepsilon_x &= \frac{du_P}{dx} = \frac{du}{dx} - y\frac{d\theta}{dx} = \frac{du}{dx} - y\frac{d^2v}{dx^2}\\ \gamma_{xy} &= \frac{dv_P}{dx} + \frac{du_P}{dy} = \frac{dv}{dx} - \theta = 0\\ \therefore \theta &= \frac{dv}{dx}\end{aligned} \tag{3.8}$$

The stress component, then takes the form

$$\begin{aligned}\sigma_x &= E\varepsilon_x = E\left(\frac{du}{dx} - y\frac{d^2v}{dx^2}\right)\\ \tau_{xy} &= 0\end{aligned} \tag{3.9}$$

where E is the elastic modulus of the beam.

We can notice that in (3.9), because there is no shear deformation consideration, only the axial strain component exists in the Euler-Bernoulli beam theory.

The equation of motion is then derived via Hamilton's principle as follows:

$$\delta H = \int_{t_1}^{t_2} (\delta S_E - \delta K_E - \delta W_E)dt = 0 \tag{3.10}$$

where

δH = the variation of total energy
δS_E = the variation of beam strain energy
δK_E = the variation of beam kinetic energy
δW_E = the variation of external work

The strain energy, S_E, of a beam element with the length L is given as

$$S_E = \frac{1}{2}\int_0^L \left(\int_A \sigma_x \varepsilon_x dA\right) dx \tag{3.11}$$

where A is the cross-sectional area of the beam.

Substitution of (3.8) and (3.9) into (3.11) gives

$$S_E = \frac{1}{2}\int_0^L \left(\int_A \left(\frac{du}{dx} - y\frac{d^2v}{dx^2}\right) E\left(\frac{du}{dx} - y\frac{d^2v}{dx^2}\right) dA\right) dx \tag{3.12}$$

Applying the variational principle to the above equation, we can have

$$\delta S_E = \int_0^L \left(\int_A E\left(\frac{du}{dx} - y\frac{d^2v}{dx^2}\right) \delta\left(\frac{du}{dx} - y\frac{d^2v}{dx^2}\right) dA\right) dx \tag{3.13}$$

For a symmetric cross section of beam with constants A and E, we have

$$\delta S_E = \int_0^L \left\{ EA\left(\frac{du}{dx}\right)\delta\left(\frac{du}{dx}\right) + EI\left(\frac{d^2v}{dx^2}\right)\delta\left(\frac{d^2v}{dx^2}\right)\right\}dx \tag{3.14}$$

where

$$A = \int_A dA = \iint dydz$$

$$I = \int_A y^2 dA = \iint y^2 dydz$$

Here, I is the second order moment of inertia of the beam cross section. Applying the integration by part to the first term of (3.14),

$$\int_0^L \left\{ EA\left(\frac{du}{dx}\right)\delta\left(\frac{du}{dx}\right)\right\}dx = EA\left(\frac{du}{dx}\right)\delta u|_{BC} - EA\int_0^L \left(\frac{d^2u}{dx^2}\right)\delta u \ dx \tag{3.15}$$

where BC is the boundary conditions of the beam, at $x = 0$ and $x = L$; after substitution the above equation becomes

$$\int_0^L \left\{ EA\left(\frac{du}{dx}\right)\delta\left(\frac{du}{dx}\right)\right\}dx = P_{x=L}\delta u_{x=L} - P_{x=0}\delta u_{x=0}$$
$$- EA\int_0^L \left(\frac{d^2u}{dx^2}\right)dx \ \delta u \tag{3.16}$$

where $P_{x=0}, P_{x=L}$ are the axial nodal forces at the beam ends which are given by

$$P_{x=0} = \left[EA\left(\frac{du}{dx}\right)\right]_{x=0}$$

$$P_{x=L} = \left[EA\left(\frac{du}{dx}\right)\right]_{x=L}$$

Applying the integration by part to the second term of (3.14),

$$\int_0^L \left\{ EI\left(\frac{d^2v}{dx^2}\right)\delta\left(\frac{d^2v}{dx^2}\right)\right\}dx = EI\left(\frac{d^2v}{dx^2}\right)\delta v|_{BC} - EI\int_0^L \left(\frac{d^3v}{dx^3}\right)dx \ \delta\left(\frac{dv}{dx}\right) \tag{3.17}$$

where BC is the boundary conditions of the beam, at $x = 0$ and $x = L$; the above equation becomes

$$\int_0^L \left\{ EI\left(\frac{d^2v}{dx^2}\right)\delta\left(\frac{d^2v}{dx^2}\right)\right\}dx = M_{x=L}\delta\left(\frac{dv}{dx}\right)_{x=L} - M_{x=0}\delta\left(\frac{dv}{dx}\right)_{x=0} - EI\int_0^L\left(\frac{d^3v}{dx^3}\right)dx\ \delta\left(\frac{dv}{dx}\right) \tag{3.18}$$

where $M_{x=0}, M_{x=L}$ are the rotational nodal moments at the beam ends which are given by

$$M_{x=0} = \left[EI\left(\frac{d^2v}{dx^2}\right)\right]_{x=0}$$

$$M_{x=L} = \left[EI\left(\frac{d^2v}{dx^2}\right)\right]_{x=L}$$

In the Euler-Bernoulli beam formulation, we notice that there is no shear strain resulted from the formulation because the shear deformation is not considered.

By further doing the integration by part to the last integral term in (3.18), we can have the following:

$$-\int_0^L \left\{ EI\left(\frac{d^3v}{dx^3}\right)\delta\left(\frac{dv}{dx}\right)\right\}dx = -Q_{x=L}\delta\left(\frac{d^3v}{dx^3}\right)_{x=L} + Q_{x=0}\delta\left(\frac{d^3v}{dx^3}\right)_{x=0} + EI\int_0^L\left(\frac{d^4v}{dx^4}\right)dx\ \delta v \tag{3.19}$$

And $Q_{x=0}, Q_{x=L}$ are the transversal nodal forces at the beam ends which are given by

$$Q_{x=0} = \left[EI\left(\frac{d^3v}{dx^3}\right)\right]_{x=0}$$

$$Q_{x=L} = \left[EI\left(\frac{d^3v}{dx^3}\right)\right]_{x=L}$$

The kinetic energy, K_E, of a beam element with the length L is given as

$$K_E = \frac{1}{2}\int_0^L\left(\int_A \rho\left(\frac{du}{dt}\right)^2 dA\right)dx + \frac{1}{2}\int_0^L\left(\int_A \rho\left(\frac{dv}{dt}\right)^2 dA\right)dx \tag{3.20}$$

where A is the cross-sectional area of the beam.

Substitution of (3.7) gives

$$K_E = \frac{1}{2}\int_0^L \left(\int_A \left(\frac{du}{dt} - y\frac{d\theta}{dt}\right)\rho\left(\frac{du}{dt} - y\frac{d\theta}{dt}\right)dA\right)dx + \frac{1}{2}\int_0^L \left(\int_A \left(\frac{dv}{dt}\right)\rho\left(\frac{dv}{dt}\right)dA\right)dx \tag{3.21}$$

Applying the variational principle to the above equation with constants A and E, then the equation becomes

$$\delta K_E = \rho A \int_0^L \left(\frac{du}{dt}\right)\ \delta\left(\frac{du}{dt}\right)dx + \rho A \int_0^L \left(\frac{dv}{dt}\right)\ \delta\left(\frac{dv}{dt}\right)dx + \rho I \int_0^L \left(\frac{d\theta}{dt}\right)\ \delta\left(\frac{d\theta}{dt}\right)dx \tag{3.22}$$

Applying the integration by part to the first term of (3.22),

$$\int_0^L \left\{\rho A\left(\frac{du}{dt}\right)\ \delta\left(\frac{du}{dt}\right)\right\}dx = \rho A\left(\frac{du}{dt}\right)\delta u|_{BC} - \rho A\int_0^L \left(\frac{d^2u}{dt^2}\right)dx\ \ \delta u \tag{3.23}$$

where BC is the boundary conditions of the beam, at $x = 0$ and $x = L$; after substitution the above equation becomes

$$\int_0^L \left\{\rho A\left(\frac{du}{dt}\right)\delta\left(\frac{du}{dt}\right)\right\}dx = P_{x=L}\delta u_{x=L} - P_{x=0}\delta u_{x=0} - \rho A\int_0^L \left(\frac{d^2u}{dt^2}\right)dx\ \ \delta u \tag{3.24}$$

and $P_{x=0}, P_{x=L}$ are the axial nodal forces at the beam ends which are given by

$$P_{x=0} = \left[\rho A\left(\frac{du}{dt}\right)\right]_{x=0}$$

$$P_{x=L} = \left[\rho A\left(\frac{du}{dt}\right)\right]_{x=L}$$

Similarly, by applying the integration by part to the second term of (3.22),

$$\int_0^L \left\{\rho A\left(\frac{dv}{dt}\right)\ \delta\left(\frac{dv}{dt}\right)\right\}dx = \rho A\left(\frac{dv}{dt}\right)\delta v|_{BC} - \rho A\int_0^L \left(\frac{d^2v}{dt^2}\right)dx\ \ \delta v \tag{3.25}$$

where BC is the boundary conditions of the beam, at $x = 0$ and $x = L$; after substitution the above equation becomes

$$\int_0^L \left\{\rho A\left(\frac{dv}{dt}\right)\delta\left(\frac{dv}{dt}\right)\right\}dx = Q_{x=L}\delta v_{x=L} - Q_{x=0}\delta v_{x=0} - \rho A\int_0^L \left(\frac{d^2v}{dt^2}\right)dx\ \delta v \quad (3.26)$$

and $Q_{x=0}, Q_{x=L}$ are the transversal nodal forces at the beam ends which are given by

$$Q_{x=0} = \left[\rho A\left(\frac{dv}{dt}\right)\right]_{x=0}$$

$$Q_{x=L} = \left[\rho A\left(\frac{dv}{dt}\right)\right]_{x=L}$$

Finally, by applying the integration by part to the third term of (3.22),

$$\int_0^L \left\{\rho I\left(\frac{d\theta}{dt}\right)\ \delta\left(\frac{d\theta}{dt}\right)\right\}dx = \rho I\left(\frac{d\theta}{dt}\right)\delta\theta|_{BC} - \rho I\int_0^L \left(\frac{d^2\theta}{dt^2}\right)dx\ \delta\theta \quad (3.27)$$

where BC is the boundary conditions of the beam, at $x = 0$ and $x = L$; after substitution the above equation becomes

$$\begin{aligned}\int_0^L \left\{\rho I\left(\frac{d\theta}{dt}\right)\delta\left(\frac{d\theta}{dt}\right)\right\}dx = \\ M_{x=L}\delta\theta_{x=L} - M_{x=0}\delta\theta_{x=0} - \rho I\int_0^L \left(\frac{d^2\theta}{dt^2}\right)dx\ \delta\theta\end{aligned} \quad (3.28)$$

and $M_{x=0}, M_{x=L}$ are the rotational nodal moments at the beam ends which are given by

$$M_{x=0} = \left[\rho I\left(\frac{d\theta}{dt}\right)\right]_{x=0}$$

$$M_{x=L} = \left[\rho I\left(\frac{d\theta}{dt}\right)\right]_{x=L}$$

Notes: In most of the textbooks, this rotational part of the beam is usually neglected due to a small influence in the Euler-Bernoulli assumptions. For completeness and consistency derivation with the Timoshenko beam, in the following section, it is included in this section.

The external work, W_E, of a beam element with the length L is given as

$$W_E = \int_0^L (p\ \ u + q\ \ v + m\ \ \theta)dx \quad (3.29)$$

where p, q, and m are the applied distributed axial force per unit length, transversal force per unit length, and moment per unit length, respectively.

Applying the variational principle to the above equation becomes

$$\delta W_E = \int_0^L (p \ \delta u + q \ \delta v + m \ \delta\theta) dx \tag{3.30}$$

In summary, by differentiating the governing equations, the differential equations of motion which correspond to generalized displacement can be summarized as

$$\begin{aligned} \delta u: \quad & -EA\frac{d^2u}{dx^2} + \rho A\frac{d^2u}{dt^2} - p = 0 \\ \delta v: \quad & +EI\frac{d^4v}{dx^4} + \rho A\frac{d^2v}{dt^2} - q = 0 \end{aligned} \tag{3.31}$$

The essential and natural boundary conditions are given as

$$\begin{aligned} \delta u: \quad & P = EA\frac{du}{dx} - \rho A\frac{du}{dt} \\ \delta v: \quad & Q = -EI\frac{d^3v}{dx^3} - \rho A\frac{dv}{dt} \\ \delta\theta: \quad & M = EI\frac{d^2v}{dx^2} - \rho I\frac{d\theta}{dt} \end{aligned} \tag{3.32}$$

3.5.2 *Straight Timoshenko Beam*

Similar with the previous section, to generalize the displacements and rotation at point P, the subindexing of the point O in (3.4) as the reference of deformation will be omitted. The assumed displacement of the Timoshenko beam can be rewritten as

$$\begin{aligned} u_P &= u - y\theta \\ v_P &= v \end{aligned} \tag{3.33}$$

From the theory of elasticity, we can obtain the strain components with shear deformation considered as

$$\begin{aligned} \varepsilon_x &= \frac{du_P}{dx} = \frac{du}{dx} - y\frac{d\theta}{dx} \\ \gamma_{xy} &= \frac{dv_P}{dx} + \frac{du_P}{dy} = \frac{dv}{dx} - \theta \\ \therefore \theta &= \frac{dv}{dx} - \gamma_{xy} \end{aligned} \tag{3.34}$$

The stress components then take the form

$$\begin{aligned} \sigma_x &= E\varepsilon_x = E\left(\frac{du}{dx} - y\frac{d\theta}{dx}\right) \\ \tau_{xy} &= G\gamma_{xy} = G\left(\frac{dv}{dx} - \theta\right) \end{aligned} \tag{3.35}$$

where G is the elastic shear modulus of the beam.

In the Timoshenko beam element, the shear deformation is considered. Hence, the strain components consist of the axial and shear strains.

The equation of motion is then derived via Hamilton's principle as follows:

$$\delta H = \int_{t_1}^{t_2} (\delta S_E - \delta K_E - \delta W_E) dt = 0 \tag{3.36}$$

where

δH = the variation of total energy
δS_E = the variation of beam strain energy
δK_E = the variation of beam kinetic energy
δW_E = the variation of external work

The strain energy, S_E, of a beam element with the length L is given as

$$S_E = \frac{1}{2}\int_0^L \left\{ \int_A (\sigma_x \varepsilon_x + \tau_{xy}\gamma_{xy}) dA \right\} dx \tag{3.37}$$

Substitution of (3.34) and (3.35) into (3.37) gives

$$S_E = \frac{1}{2}\int_0^L \left(\int_A \left\{ \begin{aligned} &\left(\frac{du}{dx} - y\frac{d\theta}{dx}\right) E \left(\frac{du}{dx} - y\frac{d\theta}{dx}\right) + \\ &\left(\frac{dv}{dx} - \theta\right) G \left(\frac{dv}{dx} - \theta\right) \end{aligned} \right\} dA \right) dx \tag{3.38}$$

Applying the variational principle to the above equation, we can have

$$\delta S_E = \int_0^L \left(\int_A \left\{ \begin{array}{l} E\left(\frac{du}{dx} - y\frac{d\theta}{dx}\right)\delta\left(\frac{du}{dx} - y\frac{d\theta}{dx}\right) + \\ G\left(\frac{dv}{dx} - \theta\right)\delta\left(\frac{dv}{dx} - \theta\right) \end{array} \right\} dA \right) dx \tag{3.39}$$

For a symmetric cross section of beam with constants A and E, the equation becomes

$$\delta S_E = \int_0^L \left\{ \begin{array}{l} EA\left(\frac{du}{dx}\right)\delta\left(\frac{du}{dx}\right) + EI\left(\frac{d\theta}{dx}\right)\delta\left(\frac{d\theta}{dx}\right) \\ +\kappa GA\left(\frac{dv}{dx} - \theta\right)\delta\left(\frac{dv}{dx} - \theta\right) \end{array} \right\} dx \tag{3.40}$$

where κ is the shear correction coefficient of the beam cross section.

Applying the integration by part to the first term of (3.40),

$$\begin{aligned} \int_0^L \left\{ EA\left(\frac{du}{dx}\right)\delta\left(\frac{du}{dx}\right) \right\} dx &= P_{x=L}\delta u_{x=L} - P_{x=0}\delta u_{x=0} \\ &- EA \int_0^L \left(\frac{d^2u}{dx^2}\right) dx\ \delta u \end{aligned} \tag{3.41}$$

and $P_{x=0}, P_{x=L}$ are the axial nodal forces at the beam ends which are given by

$$P_{x=0} = \left[EA\left(\frac{du}{dx}\right) \right]_{x=0}$$

$$P_{x=L} = \left[EA\left(\frac{du}{dx}\right) \right]_{x=L}$$

Applying the integration by part to the second term of (3.40),

$$\begin{aligned} \int_0^L \left\{ EI\left(\frac{d\theta}{dx}\right)\delta\left(\frac{d\theta}{dx}\right) \right\} dx &= M_{x=L}\delta\theta_{x=L} - M_{x=0}\delta\theta_{x=0} \\ &- EI \int_0^L \left(\frac{d^2\theta}{dx^2}\right) dx\ \delta\theta \end{aligned} \tag{3.42}$$

and $M_{x=0}, M_{x=L}$ are the rotational nodal moments at the beam ends which are given by

$$M_{x=0} = \left[EI\left(\frac{d\theta}{dx}\right) \right]_{x=0}$$

$$M_{x=L} = \left[EI\left(\frac{d\theta}{dx}\right)\right]_{x=L}$$

The integration by part to the third term of the right-hand side of (3.40) gives

$$\begin{aligned}
&\int_0^L \kappa GA\left(\frac{dv}{dx}-\theta\right)\delta\left(\frac{dv}{dx}-\theta\right)dx = \int_0^L \kappa GA\left(\frac{dv}{dx}-\theta\right)\delta\left(\frac{dv}{dx}\right)dx \\
&-\int_0^L \kappa GA\left(\frac{dv}{dx}-\theta\right)\delta\theta dx = Q_{x=L}\delta v_{x=L} - Q_{x=0}\delta v_{x=0} \\
&-\kappa GA\int_0^L \left(\frac{d^2v}{dx^2}-\frac{d\theta}{dx}\right)dx\ \delta v - \int_0^L \kappa GA\left(\frac{dv}{dx}-\theta\right)dx\ \delta\theta
\end{aligned} \tag{3.43}$$

and $Q_{x=0}, Q_{x=L}$ are the transversal nodal forces at the beam ends which are given by

$$\begin{aligned}
Q_{x=0} &= \left[\kappa GA\left(\frac{dv}{dx}-\theta\right)\right]_{x=0} = [\kappa GA\gamma]_{x=0} \\
Q_{x=L} &= \left[\kappa GA\left(\frac{dv}{dx}-\theta\right)\right]_{x=L} = [\kappa GA\gamma]_{x=L}
\end{aligned}$$

The kinetic energy, K_E, of a beam element with the length L is given as

$$K_E = \frac{1}{2}\int_0^L\left(\int_A \rho\left(\frac{du}{dt}\right)^2 dA\right)dx + \frac{1}{2}\int_0^L\left(\int_A \rho\left(\frac{dv}{dt}\right)^2 dA\right)dx \tag{3.44}$$

where A is the cross-sectional area of the beam.

Substitution of (3.33) gives

$$\begin{aligned}
K_E = &\frac{1}{2}\int_0^L\left(\int_A\left(\frac{du}{dt}-y\frac{d\theta}{dt}\right)\rho\left(\frac{du}{dt}-y\frac{d\theta}{dt}\right)dA\right)dx \\
&+\frac{1}{2}\int_0^L\left(\int_A\left(\frac{dv}{dt}\right)\rho\left(\frac{dv}{dt}\right)dA\right)dx
\end{aligned} \tag{3.45}$$

By applying the variational principle to the above equation with constant A and E, the equation becomes

$$\begin{aligned}
\delta K_E = &\rho A\int_0^L\left(\frac{du}{dt}\right)\ \delta\left(\frac{du}{dt}\right)dx + \rho A\int_0^L\left(\frac{dv}{dt}\right)\ \delta\left(\frac{dv}{dt}\right)dx \\
&+\rho I\int_0^L\left(\frac{d\theta}{dt}\right)\ \delta\left(\frac{d\theta}{dt}\right)dx
\end{aligned} \tag{3.46}$$

Applying the integration by part to the first term of (3.46),

$$\int_0^L \left\{\rho A\left(\frac{du}{dt}\right)\ \delta\left(\frac{du}{dt}\right)\right\}dx = \rho A\left(\frac{du}{dt}\right)\delta u|_{BC} - \rho A\int_0^L \left(\frac{d^2u}{dt^2}\right)dx\ \delta u \tag{3.47}$$

where BC is the boundary conditions of the beam, at $x = 0$ and $x = L$; after substitution the above equation becomes

$$\int_0^L \left\{\rho A\left(\frac{du}{dt}\right)\delta\left(\frac{du}{dt}\right)\right\}dx = P_{x=L}\delta u_{x=L} - P_{x=0}\delta u_{x=0} - \rho A\int_0^L \left(\frac{d^2u}{dt^2}\right)dx\ \delta u \tag{3.48}$$

and $P_{x=0}, P_{x=L}$ are the axial nodal forces at the beam ends which are given by

$$P_{x=0} = \left[\rho A\left(\frac{du}{dt}\right)\right]_{x=0}$$

$$P_{x=L} = \left[\rho A\left(\frac{du}{dt}\right)\right]_{x=L}$$

Similarly, by applying the integration by part to the second term of (3.46),

$$\int_0^L \left\{\rho A\left(\frac{dv}{dt}\right)\ \delta\left(\frac{dv}{dt}\right)\right\}dx = \rho A\left(\frac{dv}{dt}\right)\delta v|_{BC} - \rho A\int_0^L \left(\frac{d^2v}{dt^2}\right)dx\ \delta v \tag{3.49}$$

where BC is the boundary conditions of the beam, at $x = 0$ and $x = L$; after substitution the above equation becomes

$$\int_0^L \left\{\rho A\left(\frac{dv}{dt}\right)\delta\left(\frac{dv}{dt}\right)\right\}dx = P_{x=L}\delta v_{x=L} - P_{x=0}\delta v_{x=0} - \rho A\int_0^L \left(\frac{d^2v}{dt^2}\right)dx\ \delta v \tag{3.50}$$

and $Q_{x=0}, Q_{x=L}$ are the transversal nodal forces at the beam ends which are given by

$$Q_{x=0} = \left[\rho A\left(\frac{dv}{dt}\right)\right]_{x=0}$$

$$Q_{x=L} = \left[\rho A\left(\frac{dv}{dt}\right)\right]_{x=L}$$

Finally, by applying the integration by part to the third term of (3.46),

$$\int_0^L \left\{\rho I\left(\frac{d\theta}{dt}\right)\ \delta\left(\frac{d\theta}{dt}\right)\right\}dx = \rho I\left(\frac{d\theta}{dt}\right)\delta\theta|_{BC} - \rho I\int_0^L \left(\frac{d^2\theta}{dt^2}\right)\delta\theta\ dx \tag{3.51}$$

where BC is the boundary conditions of the beam, at $x = 0$ and $x = L$; after substitution the above equation becomes

$$\begin{aligned}\int_0^L \left\{\rho I\left(\frac{d\theta}{dt}\right)\delta\left(\frac{d\theta}{dt}\right)\right\}dx &= M_{x=L}\delta\theta_{x=L} - M_{x=0}\delta\theta_{x=0}\\ &\quad - \rho I\int_0^L \left(\frac{d^2\theta}{dt^2}\right)dx\ \delta\theta\end{aligned} \tag{3.52}$$

and $M_{x=0}, M_{x=L}$ are the rotational nodal moments at the beam ends which are given by

$$M_{x=0} = \left[\rho I\left(\frac{d\theta}{dt}\right)\right]_{x=0}$$

$$M_{x=L} = \left[\rho I\left(\frac{d\theta}{dt}\right)\right]_{x=L}$$

The external work, W_E, of a beam element with the length L is given as

$$W_E = \int_0^L (p\ \ u + q\ \ v + m\ \ \theta)dx \tag{3.53}$$

where p, q, and m are the applied distributed axial force per unit length, transversal force per unit length, and moment per unit length, respectively.

By applying the variational principle to the above equation with constant A, the equation becomes

$$\delta W_E = \int_0^L (p\ \ \delta u + q\ \ \delta v + m\ \ \delta\theta)dx \tag{3.54}$$

In summary, the equilibrium equations of the straight Timoshenko beam can be expressed by the following differential equations:

$$\begin{aligned}\delta u:&\qquad -EA\frac{d^2u}{dx^2} + \rho A\frac{d^2u}{dt^2} - p = 0\\ \delta v:&\qquad -\kappa GA\left(\frac{d^2v}{dx^2} - \frac{d\theta}{dx}\right) + \rho A\frac{d^2v}{dt^2} - q = 0\\ \delta\theta:&\qquad -EI\frac{d^2\theta}{dx^2} - \kappa GA\left(\frac{dv}{dx} - \theta\right) + \rho I\frac{d^2\theta}{dt^2} - m = 0\end{aligned} \tag{3.55}$$

The essential and natural boundary conditions are given as

$$
\begin{aligned}
\delta u: \quad & P = EA\frac{du}{dx} - \rho A\frac{du}{dt} \\
\delta v: \quad & Q = \kappa GA\left(\frac{dv}{dx} - \theta\right) - \rho A\frac{dv}{dt} \\
\delta \theta: \quad & M = EI\frac{d\theta}{dx} - \rho I\frac{d\theta}{dt}
\end{aligned} \tag{3.56}
$$

3.5.3 General Curved Euler-Bernoulli Beam

To generalize the displacements and rotation at point P, the subindexing of the point O in (3.5) as the reference of deformation will be eliminated. The assumed displacement of an arbitrary point P of the general curved Euler-Bernoulli beam can be given as

$$
\begin{aligned}
u_P &= u - r\theta - vd\varphi \\
v_P &= v - ud\varphi
\end{aligned} \tag{3.57}
$$

From the theory of elasticity, we can obtain the strain components without shear deformation as follows:

$$
\begin{aligned}
\varepsilon_s &= \frac{du_P}{ds} = \frac{du}{ds} - r\frac{d\theta}{ds} - \frac{d(vd\varphi)}{ds} = \frac{du}{ds} - r\frac{d\theta}{ds} + \frac{v}{R(s)} \\
\gamma_{sr} &= \frac{dv_P}{ds} + \frac{du_P}{dr} = \frac{dv}{ds} + \frac{u}{R(s)} - \theta = 0 \\
\therefore \theta &= \frac{dv}{ds} + \frac{u}{R(s)}
\end{aligned} \tag{3.58}
$$

where $\frac{d^2\varphi(-s)}{ds^2} = \frac{1}{R(s)}$ is the curvature of the beam at coordinate s in initial configuration.

The stress component then takes the form

$$
\begin{aligned}
\sigma_s &= E\varepsilon_s = E\left(\frac{du}{ds} - r\frac{d\theta}{ds} + \frac{v}{R(s)}\right) \\
\tau_{sr} &= 0
\end{aligned} \tag{3.59}
$$

where E is the elastic modulus of the beam.

We can notice that in (3.58), because there is no shear deformation consideration, only the axial strain component exists in the Euler-Bernoulli beam theory.

The equation of motion is then derived via Hamilton's principle as follows:

$$\delta H = \int_{t_1}^{t_2} (\delta S_E - \delta K_E - \delta W_E)dt = 0 \tag{3.60}$$

where

δH = the variation of total energy
δS_E = the variation of beam strain energy
δK_E = the variation of beam kinetic energy
δW_E = the variation of external work

The strain energy, S_E, of a beam element with the length L is given as

$$S_E = \frac{1}{2}\int_0^L \left(\int_A \sigma_s \varepsilon_s dA\right) ds \tag{3.61}$$

where A is the cross-sectional area of the beam.

Substitution of (3.58) and (3.59) into (3.61) gives

$$S_E = \frac{1}{2}\int_0^L \left(\int_A \left(\frac{du}{ds} - r\frac{d\theta}{ds} + \frac{v}{R(s)}\right) E \left(\frac{du}{ds} - r\frac{d\theta}{ds} + \frac{v}{R(s)}\right) dA\right) ds \tag{3.62}$$

Applying the variational principle to the above equation, we can have

$$\delta S_E = \int_0^L \left(\int_A E \left(\frac{du}{ds} - r\frac{d\theta}{ds} + \frac{v}{R(s)}\right) \delta \left(\frac{du}{ds} - r\frac{d\theta}{ds} + \frac{v}{R(s)}\right) dA\right) ds \tag{3.63}$$

For a symmetric cross section of beam with constants A and E,

$$\delta S_E = \int_0^L \left\{ EA \left(\frac{du}{ds} + \frac{v}{R(s)}\right) \delta \left(\frac{du}{ds} + \frac{v}{R(s)}\right) + EI \left(\frac{d\theta}{ds}\right) \delta \left(\frac{d\theta}{ds}\right) \right\} ds \tag{3.64}$$

where

$$A = \int_A dA = \iint drds$$

$$I = \int_A r^2 dA = \iint r^2 drds$$

Here, I is the second-order moment of inertia of the beam cross section.

Noting that at the initial configuration of the beam, the curvature at a coordinate s has no variation in the derivation. Applying the integration by part to the first term of (3.64),

$$\int_0^L \left\{ EA\left(\frac{du}{ds}+\frac{v}{R(s)}\right)\delta\left(\frac{du}{ds}+\frac{v}{R(s)}\right)\right\}ds = EA\left(\frac{du}{ds}+\frac{v}{R(s)}\right)\delta u|_{BC}$$
$$-EA\int_0^L\left(\frac{d^2u}{ds^2}-\frac{1}{R(s)}\frac{dv}{ds}\right)ds\ \delta u + EA\int_0^L\left(\frac{v}{R(s)^2}+\frac{1}{R(s)}\frac{du}{ds}\right)ds\ \delta v \tag{3.65}$$

where BC is the boundary conditions of the beam, at $x = 0$ and $x = L$; after substitution the above equation becomes

$$\int_0^L \left\{ EA\left(\frac{du}{ds}-\frac{v}{R(s)}\right)\delta\left(\frac{du}{ds}-\frac{v}{R(s)}\right)\right\}ds = P_{s=L}\delta u_{s=L} - P_{s=0}\delta u_{s=0}$$
$$-EA\int_0^L\left(\frac{d^2u}{ds^2}-\frac{1}{R(s)}\frac{dv}{ds}\right)ds\ \delta u + EA\int_0^L\left(\frac{v}{R(s)^2}+\frac{1}{R(s)}\frac{du}{ds}\right)ds\ \delta v \tag{3.66}$$

and $P_{s=0}, P_{s=L}$ are the axial nodal forces at the beam ends which are given by

$$P_{s=0} = \left[EA\left(\frac{du}{ds}+\frac{v}{R(s)}\right)\right]_{s=0}$$
$$P_{s=L} = \left[EA\left(\frac{du}{ds}+\frac{v}{R(s)}\right)\right]_{s=L}$$

To impose the condition of no shear deformation on the cross section of the beam, the rotation must satisfy the last equation in (3.58).

$$\theta = \frac{dv}{ds}+\frac{u}{R(s)}$$

By applying the integration by part to the second term of (3.64),

$$\int_0^L \left\{ EI\left(\frac{d\theta}{ds}\right)\delta\left(\frac{d\theta}{ds}\right)\right\}ds = \int_0^L \left\{ EI\left(\frac{d^2v}{ds^2}+\frac{1}{R(s)}\frac{du}{ds}\right)\delta\left(\frac{d^2v}{ds^2}+\frac{1}{R(s)}\frac{du}{ds}\right)\right\}ds =$$
$$+EI\left(\frac{d\theta}{ds}\right)\delta\theta|_{BC} - EI\left(\frac{1}{R(s)}\frac{d^2u}{ds^2}+\frac{d^3v}{ds^3}\right)\delta v|_{BC}$$
$$-EI\int_0^L\left(\frac{1}{R(s)}\frac{d^3v}{ds^3}+\frac{1}{R(s)^2}\frac{d^2u}{ds^2}\right)ds\ \delta u + EI\int_0^L\left(\frac{1}{R(s)}\frac{d^3u}{ds^3}+\frac{d^4v}{ds^4}\right)ds\ \delta v \tag{3.67}$$

where BC is the boundary conditions of the beam, at $s = 0$ and $s = L$, changing the term (dv/ds) to θ; after substitution the above equation becomes

$$\begin{aligned}&\int_0^L \left\{EI\left(\frac{d\theta}{ds}\right)\delta\left(\frac{d\theta}{ds}\right)\right\}ds = \int_0^L \left\{EI\left(\frac{d^2v}{ds^2} + \frac{1}{R(s)}\frac{du}{ds}\right)\delta\left(\frac{d^2v}{ds^2} + \frac{1}{R(s)}\frac{du}{ds}\right)\right\}ds = \\ &+ M_{s=L}\delta\theta_{s=L} - M_{s=0}\delta\theta_{s=0} - Q_{s=L}\delta v_{s=L} + Q_{s=0}\delta v_{s=0} \\ &- EI\int_0^L \left(\frac{1}{R(s)}\frac{d^3v}{ds^3} + \frac{1}{R(s)^2}\frac{d^2u}{ds^2}\right)ds\ \delta u + EI\int_0^L \left(\frac{1}{R(s)}\frac{d^3u}{ds^3} + \frac{d^4v}{ds^4}\right)ds\ \delta v\end{aligned} \tag{3.68}$$

and $M_{s=0}, M_{s=L}$ are the rotational nodal moments at the beam ends which are given by

$$M_{s=0} = \left[EI\left(\frac{d\theta}{ds}\right)\right]_{s=0}$$
$$M_{s=L} = \left[EI\left(\frac{d\theta}{ds}\right)\right]_{s=L}$$

and $Q_{s=0}, Q_{s=L}$ are the transversal nodal shearing forces at the beam ends which are given by

$$Q_{s=0} = \left[EI\left(\frac{1}{R(s)}\frac{d^2u}{ds^2} + \frac{d^3v}{ds^3}\right)\right]_{s=0}$$
$$Q_{s=L} = \left[EI\left(\frac{1}{R(s)}\frac{d^2u}{ds^2} + \frac{d^3v}{ds^3}\right)\right]_{s=L}$$

The kinetic energy, K_E, of a beam element with the length L is given as

$$K_E = \frac{1}{2}\int_0^L \left(\int_A \rho\left(\frac{du}{dt}\right)^2 dA\right)ds + \frac{1}{2}\int_0^L \left(\int_A \rho\left(\frac{dv}{dt}\right)^2 dA\right)ds \tag{3.69}$$

where A is the cross-sectional area of the beam.

Substitution of (3.57) gives

$$\begin{aligned}K_E = &\ \frac{1}{2}\int_0^L \left(\int_A \left(\frac{du}{dt} - r\frac{d\theta}{dt}\right)\rho\left(\frac{du}{dt} - r\frac{d\theta}{dt}\right)dA\right)ds \\ + &\ \frac{1}{2}\int_0^L \left(\int_A \left(\frac{dv}{dt}\right)\rho\left(\frac{dv}{dt}\right)dA\right)ds\end{aligned} \tag{3.70}$$

Applying the variational principle to the above equation with constant ρ, A and E, then the equations become

$$\delta K = \rho A \int_0^L \left(\frac{du}{dt}\right) \delta\left(\frac{du}{dt}\right) ds + \rho A \int_0^L \left(\frac{dv}{dt}\right) \delta\left(\frac{dv}{dt}\right) ds + \rho I \int_0^L \left(\frac{d\theta}{dt}\right) \delta\left(\frac{d\theta}{dt}\right) ds \tag{3.71}$$

Applying the integration by part to the first term of (3.71),

$$\int_0^L \left\{\rho A\left(\frac{du}{dt}\right) \delta\left(\frac{du}{dt}\right)\right\} ds = \rho A\left(\frac{du}{dt}\right)\delta u|_{BC} - \rho A \int_0^L \left(\frac{d^2u}{dt^2}\right)\delta u \; ds \tag{3.72}$$

where BC is the boundary conditions of the beam, at $s = 0$ and $s = L$; after substitution the above equation becomes

$$\int_0^L \left\{\rho A\left(\frac{du}{dt}\right)\delta\left(\frac{du}{dt}\right)\right\} ds = P_{s=L}\delta u_{s=L} - P_{s=0}\delta u_{s=0} - \rho A \int_0^L \left(\frac{d^2u}{dt^2}\right) ds \; \delta u \tag{3.73}$$

and $P_{s=0}, P_{s=L}$ are the axial nodal forces at the beam ends which are given by

$$P_{s=0} = \left[\rho A\left(\frac{du}{dt}\right)\right]_{s=0}$$

$$P_{s=L} = \left[\rho A\left(\frac{du}{dt}\right)\right]_{s=L}$$

Similarly, by applying the integration by part to the second term of (3.71),

$$\int_0^L \left\{\rho A\left(\frac{dv}{dt}\right) \delta\left(\frac{dv}{dt}\right)\right\} ds = \rho A\left(\frac{dv}{dt}\right)\delta v|_{BC} - \rho A \int_0^L \left(\frac{d^2v}{dt^2}\right)\delta v \; ds \tag{3.74}$$

where BC is the boundary conditions of the beam, at $s = 0$ and $s = L$; after substitution the above equation becomes

$$\int_0^L \left\{\rho A\left(\frac{dv}{dt}\right)\delta\left(\frac{dv}{dt}\right)\right\} ds = Q_{s=L}\delta v_{s=L} - Q_{s=0}\delta v_{s=0} - \rho A \int_0^L \left(\frac{d^2v}{dt^2}\right) ds \; \delta v \tag{3.75}$$

and $Q_{s=0}, Q_{s=L}$ are the transversal nodal shearing forces at the beam ends which are given by

$$Q_{s=0} = \left[\rho A\left(\frac{dv}{dt}\right)\right]_{s=0}$$

$$Q_{s=L} = \left[\rho A\left(\frac{dv}{dt}\right)\right]_{s=L}$$

Finally, by applying the integration by part to the third term of (3.71),

$$\int_0^L \left\{\rho I\left(\frac{d\theta}{dt}\right)\ \delta\left(\frac{d\theta}{dt}\right)\right\}ds = \rho I\left(\frac{d\theta}{dt}\right)\delta\theta|_{BC} - \rho I\int_0^L \left(\frac{d^2\theta}{dt^2}\right)\delta\theta\ ds \tag{3.76}$$

where BC is the boundary conditions of the beam, at $s = 0$ and $s = L$; after substitution the above equation becomes

$$\int_0^L \left\{\rho I\left(\frac{d\theta}{dt}\right)\delta\left(\frac{d\theta}{dt}\right)\right\}ds = M_{s=L}\delta\theta_{s=L} - M_{s=0}\delta\theta_{s=0} - \rho I\int_0^L \left(\frac{d^2\theta}{dt^2}\right)ds\,\delta\theta \tag{3.77}$$

and $M_{s=0}, M_{s=L}$ are the rotational nodal moments at the beam ends which are given by

$$M_{s=0} = \left[\rho I\left(\frac{d\theta}{dt}\right)\right]_{s=0}$$

$$M_{s=L} = \left[\rho I\left(\frac{d\theta}{dt}\right)\right]_{s=L}$$

The external work, W_E, of a beam element with the length L is given as

$$W_E = \int_0^L (p\ \ u + q\ \ v + m\ \ \theta)ds \tag{3.78}$$

Applying the variational principle to the above equation becomes

$$\delta W_E = \int_0^L (p\ \ \delta u + q\ \ \delta v + m\ \ \delta\theta)ds \tag{3.79}$$

In summary, by differentiating the governing equations, the two differential equations which correspond to generalized displacement are given as

$$\begin{aligned}
\delta u: \quad & -EA\left(\frac{d^2u}{ds^2}-\frac{1}{R(s)}\frac{dv}{ds}\right)-EI\left(\frac{1}{R(s)}\frac{d^3v}{ds^3}+\frac{1}{R(s)^2}\frac{d^2u}{ds^2}\right)\\
& +\rho A\frac{d^2u}{dt^2}-p=0\\
\delta v: \quad & EA\left(\frac{v}{R(s)^2}+\frac{1}{R(s)}\frac{du}{ds}\right)+EI\left(\frac{1}{R(s)}\frac{d^3u}{ds^3}+\frac{d^4v}{ds^4}\right)\\
& +\rho A\frac{d^2v}{dt^2}-q=0
\end{aligned} \tag{3.80}$$

The essential and natural boundary conditions are given as

$$\begin{aligned}
\delta u: \quad & P=EA\left(\frac{du}{ds}+\frac{v}{R(s)}\right)-\rho A\frac{du}{dt}\\
\delta v: \quad & Q=-EI\left(\frac{1}{R(s)}\frac{d^2u}{ds^2}+\frac{d^3v}{ds^3}\right)-\rho A\frac{dv}{dt}\\
\delta\theta: \quad & M=EI\left(\frac{d\theta}{ds}\right)-\rho I\frac{d\theta}{dt}
\end{aligned} \tag{3.81}$$

The homogeneous equilibrium equations of motion of the general curved Euler-Bernoulli beam become very complex to be solved by hand; thus, it is not easy to determine the exact shape functions like in the case of the straight Euler-Bernoulli beam. Therefore, the NURBS functions will replace the shape functions of the general curved Euler-Bernoulli beam, from which we can get the element stiffness matrix and mass matrix by using an energy method.

For circular curved Euler-Bernoulli beam element, the Eqs. (3.80) and (3.81) can be used by using a constant radius R, replacing the $R(s)$.

3.5.4 General Curved Timoshenko Beam

Similar with the previous section, to generalize the displacements and rotation at point P, the subindexing of the point O in (3.6) as the reference of deformation will be eliminated. The assumed displacement of the general curved Timoshenko beam can be given as

$$\begin{aligned}
u_P &= u - r\theta - v\,d\varphi\\
v_P &= v - u\,d\varphi
\end{aligned} \tag{3.82}$$

From the theory of elasticity, we can obtain the strain components as

$$\varepsilon_s = \frac{du_P}{ds} = \frac{du}{ds} - r\frac{d\theta}{ds} - \frac{d(vd\varphi)}{ds} = \frac{du}{ds} - r\frac{d\theta}{ds} + \frac{v}{R(s)}$$
$$\gamma_{sr} = \frac{dv_P}{ds} + \frac{du_P}{dr} = \frac{dv}{ds} - \theta + \frac{u}{R(s)} \tag{3.83}$$
$$\therefore \theta = \frac{dv}{ds} - \gamma_{sr} + \frac{u}{R(s)}$$

where $\frac{a^2\varphi(-s)}{ds^2} = \frac{1}{R(s)}$ is the curvature of the beam at coordinate s in initial configuration.

The stress components, then take the form

$$\sigma_s = E\varepsilon_s = E\left(\frac{du}{ds} - r\frac{d\theta}{ds} + \frac{v}{R(s)}\right)$$
$$\tau_{rs} = G\gamma_{rs} = G\left(\frac{dv}{ds} - \theta + \frac{u}{R(s)}\right) \tag{3.84}$$

where G is the elastic shear modulus of the beam.

In the Timoshenko beam element, the shear deformation is considered. Hence, the strain components consist of the axial and shear strains.

The equation of motion is then derived via Hamilton's principle as follows:

$$\delta H = \int_{t_1}^{t_2} (\delta S_E - \delta K_E - \delta W_E)dt = 0 \tag{3.85}$$

where

δH = the variation of total energy
δS_E = the variation of beam strain energy
δK_E = the variation of beam kinetic energy
δW_E = the variation of external work

The strain energy, S_E, of a beam element with the length L is given as

$$S_E = \frac{1}{2}\int_0^L \left\{\iint_A (\sigma_s\varepsilon_s + \tau_{sr}\gamma_{sr})dA\right\}ds \tag{3.86}$$

Substitution of (3.83) and (3.84) into (3.86) gives

$$S_E = \frac{1}{2}\int_0^L \left(\int_A \left\{\begin{array}{l} \left(\frac{du}{ds} - r\frac{d\theta}{ds} + \frac{v}{R(s)}\right)E\left(\frac{du}{ds} - r\frac{d\theta}{ds} + \frac{v}{R(s)}\right) \\ \left(+\frac{dv}{ds} - \theta + \frac{u}{R(s)}\right)G\left(\frac{dv}{ds} - \theta + \frac{u}{R(s)}\right) \end{array}\right\} dA\right) ds \tag{3.87}$$

Applying the variational principle to the above equation, we can have

$$\delta S_E = \int_0^L \left(\int_A \left\{ \begin{array}{l} \delta\left(\frac{du}{ds} - r\frac{d\theta}{ds} + \frac{v}{R(s)}\right) E\left(\frac{du}{ds} - r\frac{d\theta}{ds} + \frac{v}{R(s)}\right) \\ +\delta\left(\frac{dv}{ds} - \theta + \frac{u}{R(s)}\right) G\left(\frac{dv}{ds} - \theta + \frac{u}{R(s)}\right) \end{array} \right\} dA \right) ds \quad (3.88)$$

For vertically symmetric cross section of beam with constants A and E,

$$\delta S_E = \int_0^L \left\{ \begin{array}{l} EA\left(\frac{du}{ds} - \frac{v}{R(s)}\right)\delta\left(\frac{du}{ds} - \frac{v}{R(s)}\right) + EI\left(\frac{d\theta}{ds}\right)\delta\left(\frac{d\theta}{ds}\right) \\ +\kappa GA\left(\frac{dv}{ds} - \theta + \frac{u}{R(s)}\right)\delta\left(\frac{dv}{ds} - \theta + \frac{u}{R(s)}\right) \end{array} \right\} ds \quad (3.89)$$

where κ is the shear correction coefficient of the beam cross section.

Noting that at the initial configuration of the beam, the curvature at a coordinate s has no variation in the derivation. Applying the integration by part to the first two terms of (3.89),

$$\begin{aligned} \int_0^L \left\{ EA\left(\frac{du}{ds} + \frac{v}{R(s)}\right)\delta\left(\frac{du}{ds} + \frac{v}{R(s)}\right) \right\} ds &= P_{s=L}\delta u_{s=L} - P_{s=0}\delta u_{s=0} \\ + EA\int_0^L \left(-\frac{d^2u}{ds^2} + \frac{v}{R(s)^2} + \frac{1}{R(s)}\frac{du}{ds} - \frac{1}{R(s)}\frac{dv}{ds} \right) ds \ \delta u & \end{aligned} \quad (3.90)$$

and $P_{s=0}, P_{s=L}$ are the axial nodal forces at the beam ends which are given by

$$P_{s=0} = EA\left[\frac{du}{ds} + \frac{v}{R(s)}\right]_{s=0}$$

$$P_{s=L} = EA\left[\frac{du}{ds} + \frac{v}{R(s)}\right]_{s=L}$$

By applying the integration by part to the second term of (3.89),

$$\begin{aligned} \int_0^L \left\{ EI\left(\frac{d\theta}{ds}\right)\delta\left(\frac{d\theta}{ds}\right) \right\} ds &= M_{s=L}\delta\theta_{s=L} - M_{s=0}\delta\theta_{s=0} \\ & - EI\int_0^L \left(\frac{d^2\theta}{ds^2}\right) ds \ \delta\theta \end{aligned} \quad (3.91)$$

and $M_{s=0}, M_{s=L}$ are the rotational nodal moments at the beam ends which are given by

$$M_{s=0} = \left[EI\left(\frac{d\theta}{ds}\right)\right]_{s=0}$$

$$M_{s=L}=\left[EI\left(\frac{d\theta}{ds}\right)\right]_{s=L}$$

The integration by part to the third term of the right-hand side of (3.89) gives

$$\begin{aligned}&\int_0^L \kappa GA\left(\frac{dv}{ds}-\theta+\frac{u}{R(s)}\right)\delta\left(\frac{dv}{ds}-\theta+\frac{u}{R(s)}\right)ds=Q_{s=L}\delta v_{s=L}\\&-Q_{s=0}\delta v_{s=0}-\kappa GA\int_0^L\left(\frac{d^2v}{ds^2}-\frac{d\theta}{ds}+\frac{1}{R(s)}\frac{du}{ds}\right)ds\ \delta v\\&-\kappa GA\int_0^L\left(\frac{dv}{ds}-\theta+\frac{u}{R(s)}\right)ds\ \delta\theta+\kappa GA\int_0^L\left(\frac{1}{R(s)}\left(\frac{dv}{ds}-\theta\right)+\frac{u}{R^2(s)}\right)ds\ \delta u\end{aligned}\tag{3.92}$$

and $Q_{s=0}, Q_{s=L}$ are the transversal nodal shearing forces at the beam ends which are given by

$$Q_{s=0}=\left[\kappa GA\left(\frac{u}{R(s)}+\frac{dv}{ds}-\theta\right)\right]_{s=0}$$

$$Q_{s=L}=\left[\kappa GA\left(\frac{u}{R(s)}+\frac{dv}{ds}-\theta\right)\right]_{s=L}$$

The kinetic energy, K_E, of a beam element with the length L is given as

$$K_E=\frac{1}{2}\int_0^L\left(\int_A\rho\left(\frac{du}{dt}\right)^2dA\right)ds+\frac{1}{2}\int_0^L\left(\int_A\rho\left(\frac{dv}{dt}\right)^2dA\right)ds\tag{3.93}$$

where A is the cross-sectional area of the beam.

Substitution of (3.82) gives

$$\begin{aligned}K_E=&\frac{1}{2}\int_0^L\left(\int_A\left(\frac{du}{dt}-y\frac{d\theta}{dt}\right)\rho\left(\frac{du}{dt}-y\frac{d\theta}{dt}\right)dA\right)ds\\&+\frac{1}{2}\int_0^L\left(\int_A\left(\frac{dv}{dt}\right)\rho\left(\frac{dv}{dt}\right)dA\right)ds\end{aligned}\tag{3.94}$$

Applying the variational principle to the above equation with constants ρ, A, and E,

$$\delta K_E = \rho A \int_0^L \left(\frac{du}{dt}\right) \delta\left(\frac{du}{dt}\right) ds + \rho A \int_0^L \left(\frac{dv}{dt}\right) \delta\left(\frac{dv}{dt}\right) ds + \rho I \int_0^L \left(\frac{d\theta}{dt}\right) \delta\left(\frac{d\theta}{dt}\right) ds \tag{3.95}$$

Applying the integration by part to the first term of (3.95),

$$\int_0^L \left\{\rho A \left(\frac{du}{dt}\right) \delta\left(\frac{du}{dt}\right)\right\} ds = \rho A \left(\frac{du}{dt}\right) \delta u|_{BC} - \rho A \int_0^L \left(\frac{d^2u}{dt^2}\right) ds \ \delta u \tag{3.96}$$

where BC is the boundary conditions of the beam, at $s = 0$ and $s = L$; after substitution the above equation becomes

$$\int_0^L \left\{\rho A \left(\frac{du}{dt}\right) \delta\left(\frac{du}{dt}\right)\right\} ds = P_{s=L}\delta u_{s=L} - P_{s=0}\delta u_{s=0} - \rho A \int_0^L \left(\frac{d^2u}{dt^2}\right) ds \ \delta u \tag{3.97}$$

and $P_{s=0}, P_{s=L}$ are the axial nodal forces at the beam ends which are given by

$$P_{s=0} = \left[\rho A \left(\frac{du}{dt}\right)\right]_{s=0}$$

$$P_{s=L} = \left[\rho A \left(\frac{du}{dt}\right)\right]_{s=L}$$

Similarly, by applying the integration by part to the second term of (3.95),

$$\int_0^L \left\{\rho A \left(\frac{dv}{dt}\right) \delta\left(\frac{dv}{dt}\right)\right\} ds = \rho A \left(\frac{dv}{dt}\right) \delta v|_{BC} - \rho A \int_0^L \left(\frac{d^2v}{dt^2}\right) \delta v \ ds \tag{3.98}$$

where BC is the boundary conditions of the beam, at $s = 0$ and $s = L$; after substitution the above equation becomes

$$\int_0^L \left\{\rho A \left(\frac{dv}{dt}\right) \delta\left(\frac{dv}{dt}\right)\right\} ds = P_{s=L}\delta v_{s=L} - P_{s=0}\delta v_{s=0} - \rho A \int_0^L \left(\frac{d^2v}{dt^2}\right) ds \ \delta v \tag{3.99}$$

and $Q_{x=0}, Q_{x=L}$ are the transversal nodal shearing forces at the beam ends which are given by

$$Q_{s=0} = \left[\rho A \left(\frac{dv}{dt}\right)\right]_{s=0}$$

$$Q_{s=L} = \left[\rho A\left(\frac{dv}{dt}\right)\right]_{s=L}$$

Finally, by applying the integration by part to the third term of (3.95),

$$\int_0^L \left\{\rho I\left(\frac{d\theta}{dt}\right)\ \delta\left(\frac{d\theta}{dt}\right)\right\} ds = \rho I\left(\frac{d\theta}{dt}\right)\delta\theta|_{BC} - \rho I\int_0^L \left(\frac{d^2\theta}{dt^2}\right)\delta\theta\ ds \tag{3.100}$$

where BC is the boundary conditions of the beam, at $x = 0$ and $x = L$; after substitution the above equation becomes

$$\int_0^L \left\{\rho I\left(\frac{d\theta}{dt}\right)\delta\left(\frac{d\theta}{dt}\right)\right\} ds = P_{s=L}\delta\theta_{s=L} - P_{s=0}\delta\theta_{s=0} - \rho I\int_0^L \left(\frac{d^2\theta}{dt^2}\right) ds\ \delta\theta \tag{3.101}$$

and $M_{s=0}, M_{s=L}$ are the rotational nodal moments at the beam ends which are given by

$$M_{s=0} = \left[\rho I\left(\frac{d\theta}{dt}\right)\right]_{s=0}$$

$$M_{s=L} = \left[\rho I\left(\frac{d\theta}{dt}\right)\right]_{s=L}$$

The external work, W_E, of a beam element with the length L is given as

$$W_E = \int_0^L (p\ \ u + q\ \ v + m\ \ \theta) ds \tag{3.102}$$

Applying the variational principle to the above equation with constant A becomes

$$\delta W_E = \int_0^L (p\ \ \delta u + q\ \ \delta v + m\ \ \delta\theta) ds \tag{3.103}$$

In summary, by differentiating the governing equations, the three differential equations which correspond to generalized displacement are given as

$$
\begin{aligned}
\delta u: \quad & -EA\left(\frac{d^2u}{ds^2}+\frac{1}{R(s)}\frac{dv}{ds}\right) \\
& +\kappa GA\left(\frac{u}{R(s)^2}+\frac{dv}{ds}\frac{1}{R(s)}-\frac{\theta}{R(s)}\right)+\rho A\frac{d^2u}{dt^2}-p=0 \\
\delta v: \quad & EA\left(\frac{v}{R(s)^2}+\frac{du}{ds}\frac{1}{R(s)}\right) \\
& -\kappa GA\left(\frac{d^2v}{ds^2}+\frac{1}{R(s)}\frac{du}{ds}-\frac{d\theta}{ds}\right)+\rho A\frac{d^2v}{dt^2}-q=0 \\
\delta\theta: \quad & -EI\frac{d^2\theta}{ds^2}-\kappa GA\left(\frac{dv}{ds}+\frac{u}{R(s)}-\theta\right)+\rho I\frac{d^2\theta}{dt^2}-q=0
\end{aligned}
\tag{3.104}
$$

The essential and natural boundary conditions are given by

$$
\begin{aligned}
\delta u: \quad & P=EA\left(\frac{du}{ds}+\frac{v}{R(s)}\right)-\rho A\frac{du}{dt} \\
\delta v: \quad & Q=\kappa GA\left(\frac{u}{R(s)}+\frac{dv}{ds}-\theta\right)-\rho A\frac{dv}{dt} \\
\delta\theta: \quad & M=EI\frac{d\theta}{ds}-\rho I\frac{d\theta}{dt}
\end{aligned}
\tag{3.105}
$$

The homogeneous equilibrium equations of motion of the general curved Timoshenko beam become very complex to be solved by hand; thus, it is not easy to determine the exact shape functions like in the straight Timoshenko beam. Therefore, the NURBS functions will replace the shape functions of the general curved Timoshenko beam, from which we can get the element stiffness matrix and mass matrix by using an energy method.

For the circular curve Timoshenko beam element, Eqs. (3.104) and (3.105) can be used by using a constant radius R, replacing the $R(s)$.

3.6 Matrix Formulation of Euler-Bernoulli Beam

For the use of beam element model in the finite element analysis using a computer, the governing equations must be represented in the matrix and vector forms. In the real world, it is almost impossible to obtain analytical solutions of general beam problems. One analytical solution can only be used for a specific beam problem with predetermined boundary conditions and loadings configurations. Therefore, the numerical integration method based on the finite element method which is employed widely is adopted here.

The following sections will discuss how to obtain the Euler-Bernoulli beam formulations in the matrix and vector forms.

3.6.1 Derivation of Shape Functions

The governing equations between the axial displacement, u, and the transversal displacement, v, are uncoupled. The lowest order polynomial required is one degree for linear u to interpolate the two DOFs (u_1 and u_2) and three degreesof polynomial for v to interpolate the rest four DOFs (v_1, v_2, dv_1/dx, and dv_2/dx).

The shape functions of the Euler-Bernoulli beam can be obtained by solving the homogeneous differential equations of the static equilibrium in (3.31). By omitting the dynamic and loading terms, we can obtain two homogeneous equations as

$$\begin{aligned} EA\left(\frac{d^2u}{dx^2}\right) &= 0 \\ EI\left(\frac{d^4v}{dx^4}\right) &= 0 \end{aligned} \tag{3.106}$$

The solutions of both homogeneous equations can be approximated by the following displacement functions:

$$\begin{aligned} u &= c_0 + c_1\xi \\ v &= c_2 + c_3\xi + \frac{c_4}{2}\xi^2 + \frac{c_5}{6}\xi^3 \end{aligned} \tag{3.107}$$

where $\xi = \frac{2x-L}{L}$; $-1 \le \xi \le +1$.

The coefficients in (3.107) can be solved from the six nodal DOFs of the beam model in the finite element method from the following boundary conditions:

$$\begin{aligned} u_1 &= u_{\xi=-1} \ ; \ u_2 = u_{\xi=+1} \\ v_1 &= v_{\xi=-1} \ ; \ v_2 = v_{\xi=+1} \\ \theta_1 &= \left.\frac{dv}{dx}\right|_{\xi=-1} \ ; \ \theta_2 = \left.\frac{dv}{dx}\right|_{\xi=+1} \end{aligned} \tag{3.108}$$

By using the chain rule of $\frac{dv}{dx} = \frac{dv}{d\xi}\frac{d\xi}{dx} = \frac{2}{L}\frac{dv}{d\xi}$ and substituting (3.108) into (3.107), we can show in matrix form as

$$\begin{Bmatrix} u_1 \\ v_1 \\ \theta_1 \\ u_2 \\ v_2 \\ \theta_2 \end{Bmatrix} = \begin{bmatrix} 1 & -1 & 0 & 0 & 0 & 0 \\ 0 & 0 & -1 & -1 & 1/2 & -1/6 \\ 0 & 0 & 0 & 2/L & -2/L & 1/L \\ 1 & 1 & 0 & 0 & 0 & 0 \\ 0 & 0 & 1 & 1 & 1/2 & 1/6 \\ 0 & 0 & 0 & 2/L & 2/L & 1/L \end{bmatrix} \begin{Bmatrix} c_0 \\ c_1 \\ c_2 \\ c_3 \\ c_4 \\ c_5 \end{Bmatrix} \tag{3.109}$$

Then, by solving for the coefficient vector which are functions of nodal values, and further substituting those coefficients into (3.107), results in

$$u = \begin{bmatrix} N_{u1} & N_{u2} \end{bmatrix} \begin{Bmatrix} u_1 \\ u_2 \end{Bmatrix}$$

where

$$N_{u1} = \frac{1-\xi}{2}$$
$$N_{u2} = \frac{1+\xi}{2} \tag{3.110}$$

$$v = \begin{bmatrix} N_{vv1} & N_{v\theta 1} & N_{vv2} & N_{v\theta 2} \end{bmatrix} \begin{Bmatrix} v_1 \\ \theta_1 \\ v_2 \\ \theta_2 \end{Bmatrix}$$

where

$$N_{vv1} = \frac{1}{4}(2 - 3\xi + \xi^3)$$
$$N_{vv2} = \frac{1}{4}(2 + 3\xi - \xi^3)$$
$$N_{v\theta 1} = \frac{L}{8}(1 - \xi - \xi^2 + \xi^3)$$
$$N_{v\theta 2} = \frac{L}{8}(-1 - \xi + \xi^2 + \xi^3) \tag{3.111}$$

The shape functions for the horizontal nodal displacement, u, is shown in Fig. 3.8.

The rotational DOF, θ, can be obtained by differentiating (3.107), which results in

$$\theta = \frac{dv}{dx} = \frac{dv}{d\xi}\frac{d\xi}{dx} = \frac{2}{L}\frac{dv}{d\xi} = \frac{2}{L}\left(c_3 + c_4\xi + \frac{c_5}{2}\xi^2\right) \tag{3.112}$$

After substituting the coefficient vector of c_i of (3.109) into (3.112), we can get the rotational DOF, θ, in terms of nodal values as follows:

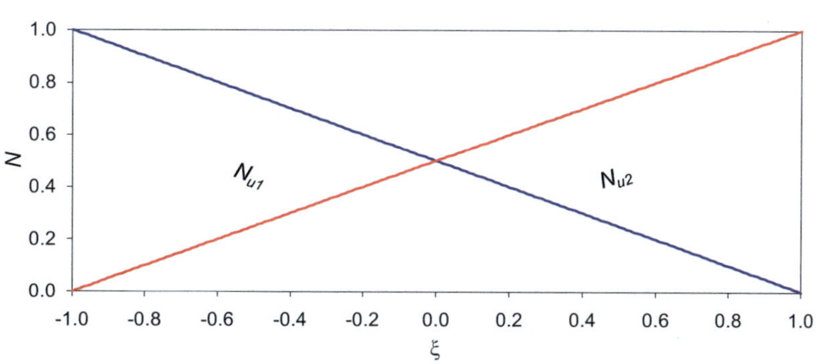

Fig. 3.8 Euler-Bernoulli nodal displacement u shape function

$$\theta = \begin{bmatrix} N_{\theta v1} & N_{\theta\theta 1} & N_{\theta v2} & N_{\theta\theta 2} \end{bmatrix} \begin{Bmatrix} v_1 \\ \theta_1 \\ v_2 \\ \theta_2 \end{Bmatrix}$$

where

$$\begin{aligned} N_{\theta v1} &= \frac{3}{2L}\left(-1+\xi^2\right) \\ N_{\theta v2} &= \frac{3}{2L}\left(1-\xi^2\right) \\ N_{\theta\theta 1} &= \frac{1}{4}\left(-1-2\xi+3\xi^2\right) \\ N_{\theta\theta 2} &= \frac{1}{4}\left(-1+2\xi+3\xi^2\right) \end{aligned} \tag{3.113}$$

The shape functions for the vertical and rotational nodal displacement, v and θ, are depicted in Fig. 3.9.

In other textbooks, the shape function is also known as displacement base functions or interpolation functions or generalized displacement. These shape functions have a role in interpolating the displacements or rotation of an arbitrary point along the beam from the nodal displacements and rotations of DOFs at both ends of the beam.

The displacements and rotation of any arbitrary point along the beam axis can be interpolated from the nodal general displacement DOF, $\mathbf{d}$, at both nodes of the beam as follows:

$$\begin{Bmatrix} u \\ v \\ \theta \end{Bmatrix} = \begin{bmatrix} N_{u1} & 0 & 0 & N_{u2} & 0 & 0 \\ 0 & N_{vv1} & N_{v\theta 1} & 0 & N_{vv2} & N_{v\theta 2} \\ 0 & N_{\theta v1} & N_{\theta\theta 1} & 0 & N_{\theta v2} & N_{\theta\theta 2} \end{bmatrix} \begin{Bmatrix} u_1 \\ v_1 \\ \theta_1 \\ u_2 \\ v_2 \\ \theta_2 \end{Bmatrix} = \mathbf{N}.\mathbf{d} \tag{3.114}$$

3.6.2 Governing Equation in Matrix Forms

The governing equation in the matrix forms can be easily obtained from the Eq. (3.10) shown below.

$$\delta H = \int_{t_1}^{t_2} (\delta S_E - \delta K_E - \delta W_E)dt = 0 \tag{3.115}$$

At an instant time of $t_1 = t_2$, the internal strain energy and kinetic energy of the beam are given as

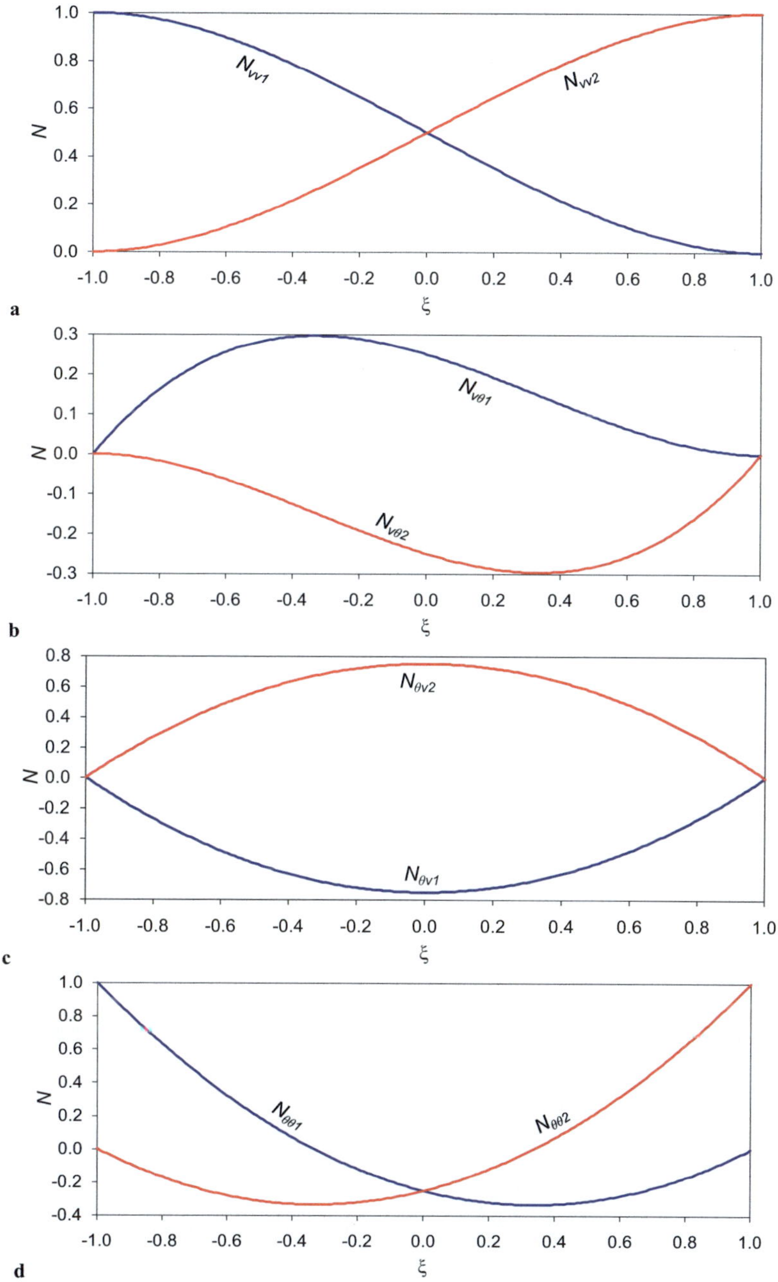

Fig. 3.9 Euler-Bernoulli beam shape function: (**a**) N_{vv1} and N_{vv2} shape functions, (**b**) $N_{v\theta 1}$ and $N_{v\theta 2}$ shape functions, (**c**) $N_{\theta v1}$ and $N_{\theta v2}$ shape functions, (**d**) $N_{\theta\theta 1}$ and $N_{\theta\theta 2}$ shape functions

$$S_E = \frac{1}{2}\int_0^L \left\{EA\left(\frac{du}{dx}\right)\left(\frac{du}{dx}\right) + EI\left(\frac{d^2v}{dx^2}\right)\left(\frac{d^2v}{dx^2}\right)\right\}dx \tag{3.116}$$

$$K_E = \frac{1}{2}\int_0^L \left\{\begin{matrix} \rho A\left(\frac{du}{dt}\right)\left(\frac{du}{dt}\right) + \rho A\left(\frac{dv}{dt}\right)\left(\frac{dv}{dt}\right) \\ +\rho I\left(\frac{d\theta}{dt}\right)\left(\frac{d\theta}{dt}\right) \end{matrix}\right\}dx \tag{3.117}$$

with the external work defined by

$$W_E = \int_0^L \{p\ \ u + q\ \ v + m\ \ \theta\}dx \tag{3.118}$$

The derivatives of displacements rotation can be represented by using the nodal general displacement vector **d** as

$$\begin{aligned} \frac{du}{dx} &= [N'_{u1} \quad N'_{u2}]\begin{Bmatrix} u_1 \\ u_2 \end{Bmatrix} \\ \frac{d^2v}{dx^2} &= \left[N''_{vv1} \quad N''_{v\theta1} \quad N''_{vv2} \quad N''_{v\theta2}\right]\begin{Bmatrix} v_1 \\ \theta_1 \\ v_2 \\ \theta_2 \end{Bmatrix} \end{aligned} \tag{3.119}$$

where

$$\begin{aligned} N'_{u1} &= -\frac{1}{L} & N'_{u2} &= \frac{1}{L} \\ N''_{v1} &= \frac{3}{2L}(\xi^2 - 1) & N''_{v2} &= \frac{3}{2L}(-\xi^2 + 1) \\ N''_{\theta1} &= \frac{1}{4}(3\xi^2 - 2\xi - 1) & N''_{\theta2} &= \frac{1}{4}(3\xi^2 + 2\xi - 1) \end{aligned}$$

By substituting (3.119) into (3.116), the internal strain energy equation in matrix forms becomes

$$S_E = \frac{1}{2}\begin{Bmatrix} u_1 \\ v_1 \\ \theta_1 \\ u_2 \\ v_2 \\ \theta_2 \end{Bmatrix}^T \int_0^L \begin{bmatrix} \mathbf{N}'_u \\ \mathbf{N}''_v \end{bmatrix}^T \begin{bmatrix} EA & 0 \\ 0 & EI \end{bmatrix}\begin{bmatrix} \mathbf{N}'_u \\ \mathbf{N}''_v \end{bmatrix} dx \begin{Bmatrix} u_1 \\ v_1 \\ \theta_1 \\ u_2 \\ v_2 \\ \theta_2 \end{Bmatrix} \tag{3.120}$$

where

$$\mathbf{N}'_u = \begin{bmatrix} N'_{u1} & 0 & 0 & N'_{u2} & 0 & 0 \end{bmatrix}$$
$$\mathbf{N}''_v = \begin{bmatrix} 0 & N''_{vv1} & N''_{v\theta 1} & 0 & N''_{vv2} & N''_{v\theta 2} \end{bmatrix}$$

By substituting (3.119) into (3.117), the kinetic energy equation in matrix forms becomes

$$K_E = \frac{1}{2} \begin{Bmatrix} \ddot{u}_1 \\ \ddot{v}_1 \\ \ddot{\theta}_1 \\ \ddot{u}_2 \\ \ddot{v}_2 \\ \ddot{\theta}_2 \end{Bmatrix}^T \int_0^L \begin{Bmatrix} \mathbf{N}'_u \\ \mathbf{N}'_v \\ \mathbf{N}'_\theta \end{Bmatrix}^T \begin{bmatrix} \rho A & 0 & 0 \\ 0 & \rho A & 0 \\ 0 & 0 & \rho I \end{bmatrix} \begin{Bmatrix} \mathbf{N}'_u \\ \mathbf{N}'_v \\ \mathbf{N}'_\theta \end{Bmatrix} dx \begin{Bmatrix} \ddot{u}_1 \\ \ddot{v}_1 \\ \ddot{\theta}_1 \\ \ddot{u}_2 \\ \ddot{v}_2 \\ \ddot{\theta}_2 \end{Bmatrix} \tag{3.121}$$

By substituting (3.119) into (3.117), the external work together with the boundary conditions, equation in matrix forms becomes

$$W_E = \begin{Bmatrix} u_1 \\ v_1 \\ \theta_1 \\ u_2 \\ v_2 \\ \theta_2 \end{Bmatrix}^T \begin{Bmatrix} N_{u1}P_1 \\ N_{v1}Q_1 \\ N_{\theta 1}M_1 \\ N_{u2}P_2 \\ N_{v2}Q_2 \\ N_{\theta 2}M_2 \end{Bmatrix} - \begin{Bmatrix} u_1 \\ v_1 \\ \theta_1 \\ u_2 \\ v_2 \\ \theta_2 \end{Bmatrix}^T \int_0^L \begin{Bmatrix} N_{u1}p \\ N_{v1}q \\ N_{\theta 1}m \\ N_{u2}p \\ N_{v2}q \\ N_{\theta 2}m \end{Bmatrix} dx \tag{3.122}$$

The theory of minimum potential energy can be used conveniently to develop the finite element formulation for a straight Euler-Bernoulli beam to obtain the equilibrium equation in matrix form.

$$\begin{aligned} \Pi &= \frac{1}{2}\mathbf{d}^T\mathbf{K}\mathbf{d} - \frac{1}{2}\ddot{\mathbf{d}}^T\mathbf{M}\ddot{\mathbf{d}} - \mathbf{d}^T\mathbf{f} \\ \frac{\partial \Pi}{\partial \mathbf{d}^T} &= \mathbf{K}\mathbf{d} - \mathbf{M}\ddot{\mathbf{d}} - \mathbf{f} = \mathbf{0} \end{aligned} \tag{3.123}$$

The stiffness, mass matrices, and loading vector are given by

$$\mathbf{K} = \int_0^L \begin{bmatrix} \mathbf{N}'_u \\ \mathbf{N}''_v \end{bmatrix}^T \begin{bmatrix} EA & 0 \\ 0 & EI \end{bmatrix} \begin{bmatrix} \mathbf{N}'_u \\ \mathbf{N}''_v \end{bmatrix} dx \tag{3.124}$$

$$\mathbf{M} = \int_0^L \begin{Bmatrix} \mathbf{N}_u \\ \mathbf{N}_v \\ \mathbf{N}_\theta \end{Bmatrix}^T \begin{bmatrix} \rho A & 0 & 0 \\ 0 & \rho A & 0 \\ 0 & 0 & \rho I \end{bmatrix} \begin{Bmatrix} \mathbf{N}_u \\ \mathbf{N}_v \\ \mathbf{N}_\theta \end{Bmatrix} dx \tag{3.125}$$

$$\mathbf{f} = \begin{Bmatrix} N_{u1}P_1 \\ N_{v1}Q_1 \\ N_{\theta 1}M_1 \\ N_{u2}P_2 \\ N_{v2}Q_2 \\ N_{\theta 2}M_2 \end{Bmatrix} + \int_0^L \begin{Bmatrix} N_{u1}p \\ N_{v1}q \\ N_{\theta 1}m \\ N_{u2}p \\ N_{v2}q \\ N_{\theta 2}m \end{Bmatrix} dx \tag{3.126}$$

3.6.3 KMmatrixEBPoly Program List

MATLAB code KMmatrixEBPoly.m and function NshapeEBPoly.m (Sect. 3.6.4) and mass matrices of a straight Euler-Bernoulli beam example given in Fig. 3.10. The computed matrices are shown in Fig. 3.11.

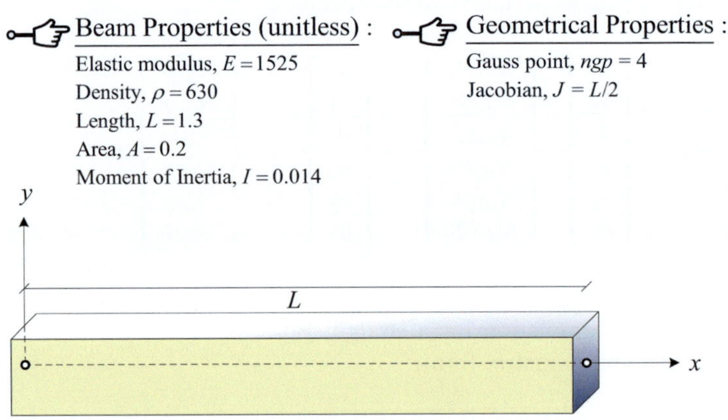

Fig. 3.10 Straight Euler-Bernoulli beam example

```
                    The Stiffness Matrix
  234.6154         0         0 -234.6154         0         0
         0  116.6136   75.7988         0 -116.6136   75.7988
         0   75.7988   65.6923         0  -75.7988   32.8462
 -234.6154         0         0  234.6154         0         0
         0 -116.6136  -75.7988         0  116.6136  -75.7988
         0   75.7988   32.8462         0  -75.7988   65.6923
```

```
                      The Mass Matrix
   54.6000         0         0   27.3000         0         0
         0   68.9815   12.0360         0   12.9185   -5.7090
         0   12.0360    4.1652         0    5.7090   -2.3595
   27.3000         0         0   54.6000         0         0
         0   12.9185    5.7090         0   68.9815  -12.0360
         0   -5.7090   -2.3595         0  -12.0360    4.1652
```

Fig. 3.11 The stiffness and mass matrices of a straight Euler-Bernoulli beam element example

```
% KMmatrixEBPoly.m
% Construct matrices K and M
% Based on the Euler-Bernoulli Beam Theory by using Polynomial
functions
clear all;clc;
L = 1.3;
E = 1525;
A = 0.2;
I = 0.014;
rho = 630;
ngp = 4;
Gpw = Legendre(ngp);
Jac = L/2;
Kmtx = zeros(6,6);
Mmtx = zeros(6,6);
Nu1 = zeros(6,ngp);
Nu0 = zeros(6,ngp);
Nv1 = zeros(6,ngp);
Nv2 = zeros(6,ngp);
Nv0 = zeros(6,ngp);
for i=1:ngp
t = Gpw(i,1);
NebP = NshapeEBPoly(0,t,Jac,L); % non derivative
Nu0(1:6,i)=NebP(1:6,1);
Nv0(1:6,i)=NebP(1:6,2);
NebP = NshapeEBPoly(1,t,Jac,L); % first derivative
Nu1(1:6,i)=NebP(1:6,1);
Nv1(1:6,i)=NebP(1:6,2);
NebP = NshapeEBPoly(2,t,Jac,L); % second derivative
Nv2(1:6,i)=NebP(1:6,2);
end
for i=1:ngp
  for j=1:6
  for k=1:6
  Kmtx(j,k) = Kmtx(j,k)+E*A*Nu1(j,i)*Nu1(k,i)*Jac*Gpw(i,2)...
              +E*I*Nv2(j,i)*Nv2(k,i)*Jac*Gpw(i,2);
  end
  end
end
% Swapping the matrices to the general displacement vector
Kmtx=Kmtx([1,3,5,2,4,6],:); % row
Kmtx=Kmtx(:,[1,3,5,2,4,6]); % column
disp('The Stiffness Matrix');
disp(Kmtx);
```

```
% Mass matrix K
for i=1:ngp
  for j=1:6
  for k=1:6
  Mmtx(j,k) = Mmtx(j,k)+rho*A*Nu0(j,i)*Nu0(k,i)*Jac*Gpw(i,2)...
 +rho*A*Nv0(j,i)*Nv0(k,i)*Jac*Gpw(i,2)...
 +rho*I*Nv1(j,i)*Nv1(k,i)*Jac*Gpw(i,2);
  end
  end
end
% Swapping the matrices to the general displacement vector
Mmtx=Mmtx([1,3,5,2,4,6],:); % row
Mmtx=Mmtx(:,[1,3,5,2,4,6]); % column
disp('The Mass Matrix');
disp(Mmtx);
```

3.6.4 NshapeEBPoly Function List

```
function NebP = NshapeEBPoly(i,t,Jac,L)
% i = 0~p derivative Shape Function, 0 not derived
% xi = Gauss coordinate
% L = beam length
%-----------------------------------
NebP=zeros(6,2); % (DOF(6), 0-2 derivatives)
switch i
case 0
 NebP(1,1) = (1-t)/2;
 NebP(2,1) = (1+t)/2;
 NebP(3,1) = 0; NebP(4,1) = 0; NebP(5,1) = 0; NebP(6,1) = 0;
 NebP(1,2) = 0; NebP(2,2) = 0;
 NebP(3,2) = 1/4*(2-3*t+t^3);
 NebP(4,2) = 1/4*(2+3*t-t^3);
 NebP(5,2) = L/8*(1-t-t^2+t^3);
 NebP(6,2) = L/8*(-1-t+t^2+t^3);
case 1
 NebP(1,1) = (-1)/2/Jac;
 NebP(2,1) = (1)/2/Jac;
 NebP(3,1) = 0; NebP(4,1) = 0; NebP(5,1) = 0; NebP(6,1) = 0;
 NebP(1,2) = 0; NebP(2,2) = 0;
 NebP(3,2) = 1/4*(-3+3*t^2)/Jac;
 NebP(4,2) = 1/4*(+3-3*t^2)/Jac;
 NebP(5,2) = L/8*(-1-2*t+3*t^2)/Jac;
 NebP(6,2) = L/8*(-1+2*t+3*t^2)/Jac;
```

```
case 2
 NebP(1,1) = 0; NebP(2,1) = 0; NebP(3,1) = 0;
 NebP(4,1) = 0; NebP(5,1) = 0; NebP(6,1) = 0;
 NebP(1,2) = 0; NebP(2,2) = 0;
 NebP(3,2) = 1/4*(6*t)/Jac/Jac;
 NebP(4,2) = 1/4*(-6*t)/Jac/Jac;
 NebP(5,2) = L/8*(-2+6*t)/Jac/Jac;
 NebP(6,2) = L/8*(2+6*t)/Jac/Jac;
end
return
```

3.7 Matrix Formulation of Timoshenko Beam

The following sections will discuss how to obtain the matrix formulation of the straight Timoshenko beam element. In the Timoshenko beam element formulation, the shear deformation is considered.

3.7.1 Derivation of Shape Functions

In a straight beam element formulation, the governing equations between the axial displacement, u, and the transversal displacement, v, have the lowest order polynomial requirement of first-degree order of linear u to interpolate the two nodal DOFs (u_1 and u_2) and third-degree order polynomial for v to interpolate the rest four nodal DOFs (v_1, v_2, θ_1, and θ_2).

The shape functions of the Timoshenko beam can be obtained by solving the homogeneous differential equations of the static equilibrium in (3.55). The shape functions can be obtained by neglecting the dynamic and loading terms in the equilibrium equations as given below.

$$\begin{aligned} & EA\left(\frac{d^2u}{dx^2}\right) = 0 \\ & \frac{dv}{dx}\left(\kappa GA\left(\frac{dv}{dx} - \theta\right)\right) = 0 \\ & EI\left(\frac{d^2\theta}{dx^2}\right) + \kappa GA\left(\frac{dv}{dx} - \theta\right) = 0 \end{aligned} \tag{3.127}$$

The solutions of the first equation in (3.127) can be obtained by integrating the differential equation. The linear polynomial function can be assumed to solve the first differential equation of horizontal nodal displacement u as follows:

$$EA\left(\frac{du}{dx}\right) = c_0 \tag{3.128}$$

Integrating (3.128), by noting that $\xi = \frac{2x-L}{L}$; $-1 \leq \xi \leq +1$, and using the chain rule of $\frac{du}{dx} = \frac{du}{d\xi}\frac{d\xi}{dx} = \frac{2}{L}\frac{du}{d\xi}$ result in

$$u = c_1 + \frac{c_0 L^2}{4EA}\xi \tag{3.129}$$

The remaining second and third equations of (3.127) have to be solved together because they are coupled. The second equation implies that the term in the bracket of should be a constant value to have its derivative of zero.

$$\kappa GA\left(\frac{dv}{dx} - \theta\right) = c_2 \tag{3.130}$$

where c_2 is a constant. Substituting the above relationship to the third equation, we can have the following equation:

$$EI\left(\frac{d^2\theta}{dx^2}\right) + c_2 = 0 \tag{3.131}$$

Integrating two times the (3.131), by noting that $\xi = \frac{2x-L}{L}$; $-1 \leq \xi \leq +1$, and using the chain rule of $\frac{dv}{dx} = \frac{dv}{d\xi}\frac{d\xi}{dx} = \frac{2}{L}\frac{dv}{d\xi}$; $\frac{d\theta}{dx} = \frac{d\theta}{d\xi}\frac{d\xi}{dx} = \frac{2}{L}\frac{d\theta}{d\xi}$, result in

$$\theta = c_4 + c_3\xi - \frac{c_2 L^2}{8EI}\xi^2 \tag{3.132}$$

Now, substituting back the θ into the second equation, we can get the following after integrating the v with respect to x:

$$v = c_5 + \left(\frac{c_2 L}{2\kappa GA} + \frac{c_4 L}{2}\right)\xi + \frac{c_3 L}{4}\xi^2 - \frac{c_2 L^3}{48EI}\xi^3 \tag{3.133}$$

Now, we have six unknowns that can be solved by using the six boundary conditions at both ends of the beam defined as follows:

$$\begin{aligned} u_1 &= u_{\xi=-1} \ ; \ u_2 = u_{\xi=+1} \\ v_1 &= v_{\xi=-1} \ ; \ v_2 = v_{\xi=+1} \\ \theta_1 &= \theta_{\xi=-1} \ ; \ \theta_2 = \theta_{\xi=+1} \end{aligned} \tag{3.134}$$

By substituting (3.134) into (3.129), (3.132), and (3.133), expressed in the matrix form as follows,

$$\begin{Bmatrix} u_1 \\ w_1 \\ \theta_1 \\ u_2 \\ w_2 \\ \theta_2 \end{Bmatrix} = \begin{bmatrix} -\frac{L^2}{4EA} & 1 & 0 & 0 & 0 & 0 \\ 0 & 0 & \left(-\frac{L}{2\kappa GA}+\frac{L^3}{48EI}\right) & \frac{L}{4} & -\frac{L}{2} & 1 \\ 0 & 0 & -\frac{L^2}{8EI} & -1 & 1 & 0 \\ \frac{L^2}{4EA} & 1 & 0 & 0 & 0 & 0 \\ 0 & 0 & \left(\frac{L}{2\kappa GA}-\frac{L^3}{48EI}\right) & \frac{L}{4} & \frac{L}{2} & 1 \\ 0 & 0 & \frac{L^2}{8EI} & 1 & 1 & 0 \end{bmatrix} \begin{Bmatrix} c_1 \\ c_2 \\ c_3 \\ c_4 \\ c_5 \\ c_6 \end{Bmatrix} \tag{3.135}$$

and solving for the coefficient vector, further substituting all the coefficients into (3.129), (3.132), and (3.133), result in

$$\begin{aligned} &u = \begin{bmatrix} N_{u1} & N_{u2} \end{bmatrix} \begin{Bmatrix} u_1 \\ u_2 \end{Bmatrix} \\ &\text{where} \\ &N_{u1} = \frac{1-\xi}{2} \\ &N_{u2} = \frac{1+\xi}{2} \end{aligned} \tag{3.136}$$

defining the ratio of the beam bending to shear stiffness as follows:

$$\phi = \frac{12}{L^2}\left(\frac{EI}{\kappa GA}\right) \tag{3.137}$$

The shape functions for v and θ which are the function of ξ and ϕ are given as

$$\begin{aligned} &v = \begin{bmatrix} N_{vv1} & N_{v\theta 1} & N_{vv2} & N_{v\theta 2} \end{bmatrix} \begin{Bmatrix} v_1 \\ \theta_1 \\ v_2 \\ \theta_2 \end{Bmatrix} \\ &\text{where} \\ &N_{vv1} = \frac{1}{4(1+\phi)}\left\{(\xi-1)\left(\xi^3+\xi-2(1+\phi)\right)\right\} \\ &N_{v\theta 1} = \frac{L}{8(1+\phi)}\left\{\left(\xi^2-1\right)(\xi-(1+\phi))\right\} \\ &N_{vv2} = \frac{1}{4(1+\phi)}\left\{(\xi+1)\left(-\xi^3+\xi+2(1+\phi)\right)\right\} \\ &N_{v\theta 2} = \frac{L}{8(1+\phi)}\left\{\left(\xi^2-1\right)(\xi+(1+\phi))\right\} \end{aligned} \tag{3.138}$$

$$\theta = [N_{\theta v1} \quad N_{\theta\theta 1} \quad N_{\theta v2} \quad N_{\theta\theta 2}] \begin{Bmatrix} v_1 \\ \theta_1 \\ v_2 \\ \theta_2 \end{Bmatrix}$$

where

$$\begin{aligned} N_{\theta v1} &= \frac{3}{2L(1+\phi)}(\xi^2 - 1) \\ N_{\theta\theta 1} &= \frac{1}{4(1+\phi)}\{(\xi - 1)(3\xi + 1 - 2\phi)\} \\ N_{\theta v2} &= \frac{3}{2L(1+\phi)}(-\xi^2 + 1) \\ N_{\theta\theta 2} &= \frac{1}{4(1+\phi)}\{(\xi + 1)(3\xi - 1 + 2\phi)\} \end{aligned} \tag{3.139}$$

These shape functions have a role in interpolating the displacements or rotation of an arbitrary point along the beam from the nodal displacements and rotations of DOFs at both ends of the beam. The shape functions for the horizontal nodal displacement u are shown in Fig. 3.8; vertical and rotational v and θ are shown in Fig. 3.12 for different values of $\phi = 0$, 1, and 10.

It should be noted that in the shape function, the ratio, ϕ, between the bending and shear stiffness of the beam is taken into consideration. It can be shown that the stiffness and mass matrices of the Timoshenko straight beam element that will be constructed later will reduce to the Euler-Bernoulli beam element by setting $\phi = 0$.

3.7.2 Governing Equation in Matrix Forms

The governing equation in the matrix forms can be obtained from the Eq. (3.36).

$$\delta H = \int_{t_1}^{t_2} (\delta S_E - \delta K_E - \delta W_E)dt = 0 \tag{3.140}$$

At an instant time of $t_1 = t_2$, the internal strain and kinetic energy of the beam are given as

$$S_E = \frac{1}{2}\int_0^L \left\{ \begin{array}{l} EA\left(\frac{du}{dx}\right)\left(\frac{du}{dx}\right) + EI\left(\frac{d\theta}{dx}\right)\left(\frac{d\theta}{dx}\right) \\ +\kappa GA\left(\frac{dv}{dx} - \theta\right)\left(\frac{dv}{dx} - \theta\right) \end{array} \right\} dx \tag{3.141}$$

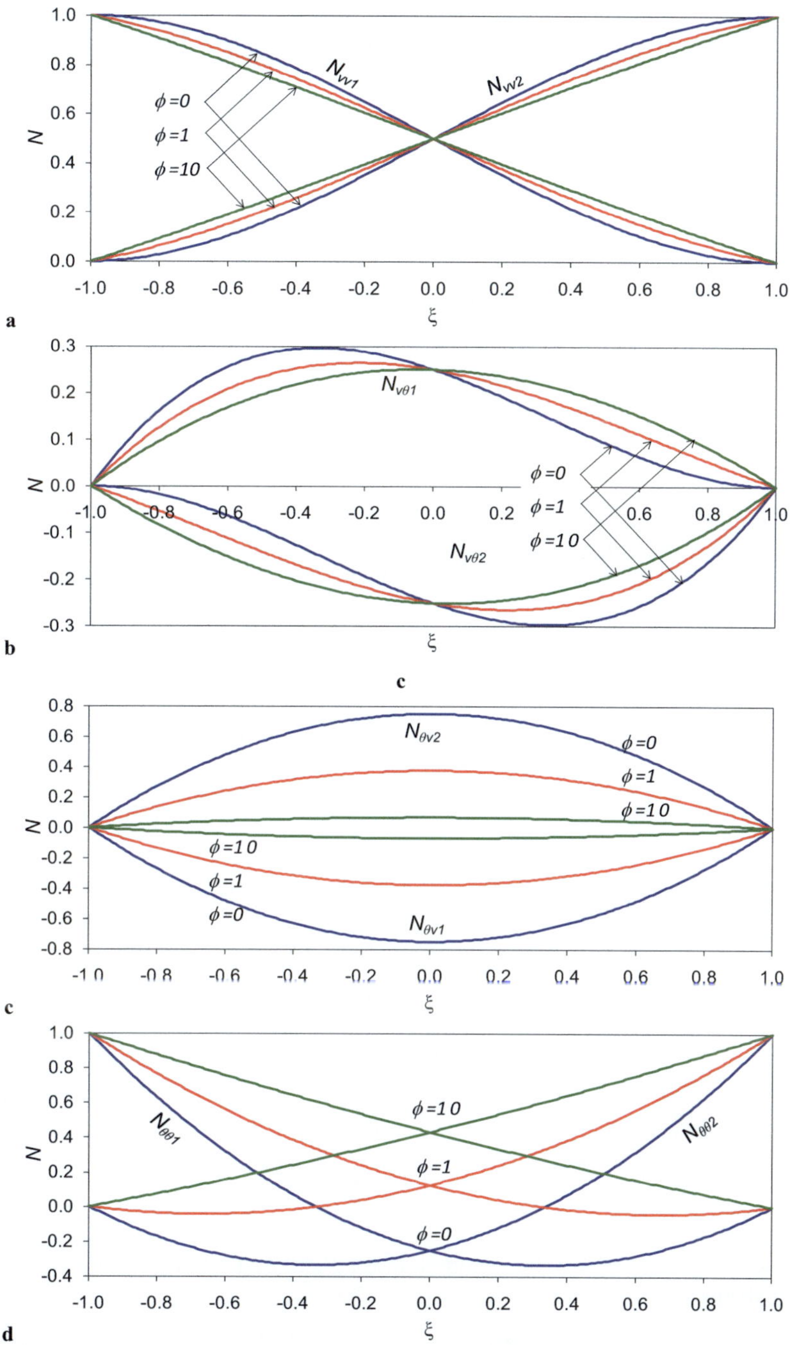

Fig. 3.12 Timoshenko beam shape function with various ϕ values: (**a**) N_{vv1} and N_{vv2} shape functions, (**b**) $N_{v\theta 1}$ and $N_{v\theta 2}$ shape functions, (**c**) $N_{\theta v1}$ and $N_{\theta v2}$ shape functions, (**d**) $N_{\theta\theta 1}$ and $N_{\theta\theta 2}$ shape functions

$$K_E = \frac{1}{2}\int_0^L \left\{ \begin{array}{l} \rho A\left(\frac{du}{dt}\right)\left(\frac{du}{dt}\right) + \rho A\left(\frac{dv}{dt}\right)\left(\frac{dv}{dt}\right) \\ +\rho I\left(\frac{d\theta}{dt}\right)\left(\frac{d\theta}{dt}\right) \end{array} \right\} dx \tag{3.142}$$

with the external work defined by

$$W_E = \int_0^L \{p\ u + q\ v + m\ \theta\} dx \tag{3.143}$$

The derivatives of displacements and rotations can be represented by using the nodal general displacement vector **d** as

$$\begin{aligned} \frac{du}{dx} &= [N'_{u1} \quad N'_{u2}]\left\{\begin{array}{c} u_1 \\ u_2 \end{array}\right\} \\ \frac{dv}{dx} &= \left[N'_{vv1} \quad N'_{v\theta 1} \quad N'_{vv2} \quad N'_{v\theta 2}\right]\left\{\begin{array}{c} v_1 \\ v_2 \\ \theta_1 \\ \theta_2 \end{array}\right\} \\ \frac{d\theta}{dx} &= \left[N'_{\theta v1} \quad N'_{\theta\theta 1} \quad N'_{\theta v2} \quad N'_{\theta\theta 2}\right]\left\{\begin{array}{c} v_1 \\ v_2 \\ \theta_1 \\ \theta_2 \end{array}\right\} \end{aligned} \tag{3.144}$$

where

$$\begin{aligned} N'_{u1} &= -\frac{1}{L} & N'_{u2} &= \frac{1}{L} \\ N'_{vv1} &= \frac{1}{4}(-3+\xi^2) & N'_{vv2} &= \frac{1}{4}(3-\xi^2) \\ N'_{v\theta 1} &= \frac{L}{8}(-1-2\xi+3\xi^2) & N'_{v\theta 2} &= \frac{L}{8}(-1+2\xi+3\xi^2) \\ N'_{\theta v1} &= \frac{3\xi}{L} & N'_{\theta v2} &= -\frac{3\xi}{L} \\ N'_{\theta\theta 1} &= \frac{(3\xi-1)}{2} & N'_{\theta\theta 1} &= \frac{(3\xi+1)}{2} \end{aligned}$$

By substituting (3.144), the internal strain energy equation can be expressed in matrix forms as

$$S_E = \frac{1}{2}\begin{Bmatrix} u_1 \\ v_1 \\ \theta_1 \\ u_2 \\ v_2 \\ \theta_2 \end{Bmatrix}^T \int_0^L \begin{Bmatrix} \begin{Bmatrix} \mathbf{N}'_u \\ (\mathbf{N}'_v - \mathbf{N}_\theta) \\ \mathbf{N}'_\theta \end{Bmatrix}^T \begin{bmatrix} EA & 0 & 0 \\ 0 & \kappa GA & 0 \\ 0 & 0 & EI \end{bmatrix} \\ \times \begin{Bmatrix} \mathbf{N}'_u \\ (\mathbf{N}'_v - \mathbf{N}_\theta) \\ \mathbf{N}'_\theta \end{Bmatrix} \end{Bmatrix} dx \begin{Bmatrix} u_1 \\ v_1 \\ \theta_1 \\ u_2 \\ v_2 \\ \theta_2 \end{Bmatrix} \tag{3.145}$$

where

$$\begin{aligned} \mathbf{N}'_u &= \begin{bmatrix} N'_{u1} & 0 & 0 & N'_{u2} & 0 & 0 \end{bmatrix} \\ \mathbf{N}'_v &= \begin{bmatrix} 0 & N'_{vv1} & N'_{v\theta1} & 0 & N'_{vv2} & N'_{v\theta2} \end{bmatrix} \\ \mathbf{N}'_\theta &= \begin{bmatrix} 0 & N'_{\theta v1} & N'_{\theta\theta1} & 0 & N'_{\theta v2} & N'_{\theta\theta2} \end{bmatrix} \end{aligned}$$

By substituting (3.144), the kinetic energy equation can be expressed in matrix forms as

$$K_E = \begin{Bmatrix} \ddot{u}_1 \\ \ddot{v}_1 \\ \ddot{\theta}_1 \\ \ddot{u}_2 \\ \ddot{v}_2 \\ \ddot{\theta}_2 \end{Bmatrix}^T \frac{1}{2}\int_0^L \begin{Bmatrix} \mathbf{N}'_u \\ \mathbf{N}'_v \\ \mathbf{N}_\theta \end{Bmatrix}^T \begin{bmatrix} \rho A & 0 & 0 \\ 0 & \rho A & 0 \\ 0 & 0 & \rho I \end{bmatrix} \begin{Bmatrix} \mathbf{N}'_u \\ \mathbf{N}'_v \\ \mathbf{N}_\theta \end{Bmatrix} dx \begin{Bmatrix} \ddot{u}_1 \\ \ddot{v}_1 \\ \ddot{\theta}_1 \\ \ddot{u}_2 \\ \ddot{v}_2 \\ \ddot{\theta}_2 \end{Bmatrix} \tag{3.146}$$

By substituting (3.144) into (3.143), the external work together with the boundary conditions equation in matrix form becomes

$$W_E = \begin{Bmatrix} u_1 \\ v_1 \\ \theta_1 \\ u_2 \\ v_2 \\ \theta_2 \end{Bmatrix}^T \begin{Bmatrix} N_{u1}P_1 \\ N_{v1}Q_1 \\ N_{\theta1}M_1 \\ N_{u2}P_2 \\ N_{v2}Q_2 \\ N_{\theta2}M_2 \end{Bmatrix} - \begin{Bmatrix} u_1 \\ v_1 \\ \theta_1 \\ u_2 \\ v_2 \\ \theta_2 \end{Bmatrix}^T \int_0^L \begin{Bmatrix} N_{u1}p \\ N_{v1}q \\ N_{\theta1}m \\ N_{u2}p \\ N_{v2}q \\ N_{\theta2}m \end{Bmatrix} dx \tag{3.147}$$

The theory of minimum potential energy can be used conveniently to develop the finite element formulation for a straight Timoshenko beam to obtain the equilibrium equation in matrix form.

$$\begin{aligned} \Pi &= \frac{1}{2}\mathbf{d}^T\mathbf{K}\mathbf{d} - \frac{1}{2}\ddot{\mathbf{d}}^T\mathbf{M}\ddot{\mathbf{d}} - \mathbf{d}^T\mathbf{f} \\ \frac{\partial \Pi}{\partial \mathbf{d}^T} &= \mathbf{K}\mathbf{d} - \mathbf{M}\ddot{\mathbf{d}} - \mathbf{f} = \mathbf{0} \end{aligned} \tag{3.148}$$

The stiffness matrix, mass matrix, and loading vector are given by

$$\mathbf{K} = \int_0^L \left\{ \begin{Bmatrix} \mathbf{N}'_u \\ (\mathbf{N}'_v - \mathbf{N}_\theta) \\ \mathbf{N}'_\theta \end{Bmatrix}^T \begin{bmatrix} EA & 0 & 0 \\ 0 & \kappa GA & 0 \\ 0 & 0 & EI \end{bmatrix} \begin{Bmatrix} \mathbf{N}'_u \\ (\mathbf{N}'_v - \mathbf{N}_\theta) \\ \mathbf{N}'_\theta \end{Bmatrix} \right\} dx \tag{3.149}$$

$$\mathbf{M} = \int_0^L \begin{Bmatrix} \mathbf{N}_u \\ \mathbf{N}_v \\ \mathbf{N}_\theta \end{Bmatrix}^T \begin{bmatrix} \rho A & 0 & 0 \\ 0 & \rho A & 0 \\ 0 & 0 & \rho I \end{bmatrix} \begin{Bmatrix} \mathbf{N}_u \\ \mathbf{N}_v \\ \mathbf{N}_\theta \end{Bmatrix} dx \tag{3.150}$$

$$\mathbf{f} = \begin{Bmatrix} N_{u1}P_1 \\ N_{v1}Q_1 \\ N_{\theta1}M_1 \\ N_{u2}P_2 \\ N_{v2}Q_2 \\ N_{\theta2}M_2 \end{Bmatrix} + \int_0^L \begin{Bmatrix} N_{u1}p \\ N_{v1}q \\ N_{\theta1}m \\ N_{u2}p \\ N_{v2}q \\ N_{\theta2}m \end{Bmatrix} dx \tag{3.151}$$

3.7.3 KMmatrixTiPoly Program List

MATLAB code KMmatrixTiPoly.m and function NshapeTiPoly.m (Sect. 3.7.4) compute the stiffness and mass matrices of a straight Timoshenko beam element example which is shown in Fig. 3.13. The computed matrices are given in Fig. 3.14.

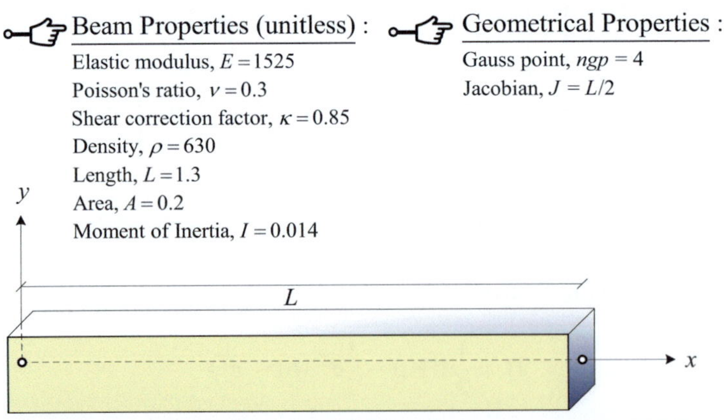

Fig. 3.13 Straight Timoshenko beam example

The Stiffness Matrix

234.6154	0	0	-234.6154	0	0
0	46.2686	30.0746	0	-46.2686	30.0746
0	30.0746	35.9716	0	-30.0746	3.1254
-234.6154	0	0	234.6154	0	0
0	-46.2686	-30.0746	0	46.2686	-30.0746
0	30.0746	3.1254	0	-30.0746	35.9716

The Mass Matrix

54.6000	0	0	27.3000	0	0
0	58.1708	8.7397	0	23.7292	-9.0053
0	8.7397	4.4476	0	9.0053	-2.0771
27.3000	0	0	54.6000	0	0
0	23.7292	9.0053	0	58.1708	-8.7397
0	-9.0053	-2.0771	0	-8.7397	4.4476

Fig. 3.14 The stiffness and mass matrices of a straight Timoshenko beam element example

```
% KMmatrixTiPoly.m
% Construct matrices K and M
% Based on the Timoshenko Beam Theory by using Polynomial func-
tions
clear all;clc;
L = 1.3;
E = 1525;
A = 0.2;
kp = 0.85;
I = 0.014;
nu = 0.3;
G = E/2/(1+nu);
phi = 12/L/L*E*I/kp/G/A;
rho = 630;
ngp = 4;
Gpw = Legendre(ngp);
Jac = L/2;
Kmtx = zeros(6,6);
Mmtx = zeros(6,6);
Nu1 = zeros(6,ngp);
Nv1 = zeros(6,ngp);
Nt1 = zeros(6,ngp);
Nu0 = zeros(6,ngp);
Nv0 = zeros(6,ngp);
Nt0 = zeros(6,ngp);
for i=1:ngp
t = Gpw(i,1);
NTiP = NshapeTiPoly(0,t,Jac,L,phi); % non derivative
Nu0(1:6,i)=NTiP(1:6,1);
```

```
Nv0(1:6,i)=NTiP(1:6,2);
Nt0(1:6,i)=NTiP(1:6,3);
NTiP = NshapeTiPoly(1,t,Jac,L,phi); % first derivative
Nu1(1:6,i)=NTiP(1:6,1);
Nv1(1:6,i)=NTiP(1:6,2);
Nt1(1:6,i)=NTiP(1:6,3);
end
for i=1:ngp
  for j=1:6
    for k=1:6
      Kmtx(j,k) = Kmtx(j,k)+E*A*Nu1(j,i)*Nu1(k,i)*Jac*Gpw(i,2)...
      +E*I*Nt1(j,i)*Nt1(k,i)*Jac*Gpw(i,2)+kp*G*A*(Nv1(j,i)-Nt0(j,
      i))...
      *(Nv1(k,i)-Nt0(k,i))*Jac*Gpw(i,2);
    end
  end
end
% Swapping the matrices to the general displacement vector
Kmtx=Kmtx([1,3,5,2,4,6],:); % row
Kmtx=Kmtx(:,[1,3,5,2,4,6]); % column
disp('The Stiffness Matrix');
disp(Kmtx);
% Mass matrix K
for i=1:ngp
  for j=1:6
    for k=1:6
      Mmtx(j,k) = Mmtx(j,k)+rho*A*Nu0(j,i)*Nu0(k,i)*Jac*Gpw(i,2)...
      + rho*A*Nv0(j,i)*Nv0(k,i)*Jac*Gpw(i,2)...
      + rho*I*Nt0(j,i)*Nt0(k,i)*Jac*Gpw(i,2);
    end
  end
end
% Swapping the matrices to the general displacement vector
Mmtx=Mmtx([1,3,5,2,4,6],:); % row
Mmtx=Mmtx(:,[1,3,5,2,4,6]); % column
disp('The Mass Matrix');
disp(Mmtx);
```

3.7.4 NshapeTiPoly Function List

```
function NTiP = NshapeTiPoly(i,t,Jac,L,phi)
% i = 0~p derivative Shape Function, 0 not derived
% xi = Gauss coordinate
```

```
% L = beam length
%------------------------------------
NTiP=zeros(6,3); % (DOF(6), 0-1 derivatives)
switch i
case 0
 NTiP(1,1) = (1-t)/2;
 NTiP(2,1) = (1+t)/2;
 NTiP(3,1) = 0; NTiP(4,1) = 0; NTiP(5,1) = 0; NTiP(6,1) = 0;
 NTiP(1,2) = 0; NTiP(2,2) = 0;
 NTiP(3,2) = 1/4/(1+phi)*(t-1)*(t^2+t-2*(1+phi));
 NTiP(4,2) = 1/4/(1+phi)*(2*(t+1)*(1+phi)+t-t^3);
 NTiP(5,2) = L/8/(1+phi)*((t^2-1)*(t-1-phi));
 NTiP(6,2) = L/8/(1+phi)*((t^2-1)*(t+1+phi));
 NTiP(1,3) = 0; NTiP(2,3) = 0;
 NTiP(3,3) = 3/2/L/(1+phi)*(t^2-1);
 NTiP(4,3) = 3/2/L/(1+phi)*(-t^2+1);
 NTiP(5,3) = 1/4/(1+phi)*((t-1)*(3*t+1-2*phi));
 NTiP(6,3) = 1/4/(1+phi)*((t+1)*(3*t-1+2*phi));
case 1
 NTiP(1,1) = (-1)/2/Jac;
 NTiP(2,1) = (1)/2/Jac;
 NTiP(3,1) = 0; NTiP(4,1) = 0; NTiP(5,1) = 0; NTiP(6,1) = 0;
 NTiP(1,2) = 0; NTiP(2,2) = 0;
 NTiP(3,2) = 1/4/(1+phi)*(-3+3*t^2-2*phi)/Jac;
 NTiP(4,2) = 1/4/(1+phi)*(3-3*t^2+2*phi)/Jac;
 NTiP(5,2) = L/8/(1+phi)*(-1+3*t^2-2*t*(1+phi))/Jac;
 NTiP(6,2) = L/8/(1+phi)*(-1+3*t^2+2*t*(1+phi))/Jac;
 NTiP(1,3) = 0; NTiP(2,3) = 0;
 NTiP(3,3) = 3/L/(1+phi)*t/Jac;
 NTiP(4,3) = -3/L/(1+phi)*t/Jac;
 NTiP(5,3) = 1/2/(1+phi)*(3*t-1-phi)/Jac;
 NTiP(6,3) = 1/2/(1+phi)*(3*t+1+phi)/Jac;
end
return
```

3.8 Matrix Formulation of General Curved Euler-Bernoulli Beam

Unlike the straight beam element discussed in the previous section, in this section, the derivation of shape functions will not be presented because of the complexity of general curves which makes it difficult to find a single analytical solution by solving the homogeneous differential equations of (3.80) and compatibility displacements due to the enforcing the condition of no shear deformation of the beam cross section $\theta = dv/ds + u/R(s)$. However, with the help of NURBS that will be discussed in the following chapter, the shape functions will be replaced by the NURBS functions that can be integrated numerically.

3.8.1 Governing Equation in Matrix Forms

The governing equation in the matrix forms can be obtained from the Eq. (3.60).

$$\delta H = \int_{t_1}^{t_2} (\delta S_E - \delta K_E - \delta W_E) dt = 0 \tag{3.152}$$

At an instant time of $t_1 = t_2$, the internal strain and kinetic energy of the beam are given as

$$\delta S_E = \int_0^L \left\{ EA\left(\frac{du}{ds} + \frac{v}{R(s)}\right)\delta\left(\frac{du}{ds} + \frac{v}{R(s)}\right) + EI\left(\frac{d\theta}{ds}\right)\delta\left(\frac{d\theta}{ds}\right) \right\} ds \tag{3.153}$$

$$\begin{aligned} K_E = {} & \frac{1}{2}\int_0^L \left(\int_A \left(\frac{du}{dt} - y\frac{d\theta}{dt}\right)\rho\left(\frac{du}{dt} - y\frac{d\theta}{dt}\right) dA \right) ds \\ & + \frac{1}{2}\int_0^L \left(\int_A \left(\frac{dv}{dt}\right)\rho\left(\frac{dv}{dt}\right) dA \right) ds \end{aligned} \tag{3.154}$$

with the external work defined by

$$W_E = \int_0^L \{p\ u + q\ v + m\ \theta\} dx \tag{3.155}$$

The derivatives of displacements and rotations can be represented by using the nodal general displacement vector $\mathbf{d} = \{u_1 \quad v_1 \quad \theta_1 \quad u_2 \quad v_2 \quad \theta_2\}^T$ as given by

$$\begin{aligned} \frac{du}{ds} &= \begin{bmatrix} N'_{uu1} & N'_{uv1} & N'_{u\theta 1} & N'_{uu2} & N'_{uv2} & N'_{u\theta 2} \end{bmatrix} \mathbf{d} \\ \frac{dv}{ds} &= \begin{bmatrix} N'_{vu1} & N'_{vv1} & N'_{v\theta 1} & N'_{vu2} & N'_{vv2} & N'_{v\theta 2} \end{bmatrix} \mathbf{d} \\ \frac{d\theta}{ds} &= \begin{bmatrix} N'_{\theta u1} & N'_{\theta v1} & N'_{\theta\theta 1} & N'_{\theta u2} & N'_{\theta v2} & N'_{\theta\theta 2} \end{bmatrix} \mathbf{d} \end{aligned} \tag{3.156}$$

By substituting (3.156), the internal strain energy equation can be expressed in matrix forms as

$$S_E = \frac{1}{2}\begin{Bmatrix} u_1 \\ v_1 \\ \theta_1 \\ u_2 \\ v_2 \\ \theta_2 \end{Bmatrix}^T \int_0^L \begin{Bmatrix} \mathbf{N}'_u + \frac{\mathbf{N}_v}{R(s)} \\ \mathbf{N}'_\theta \end{Bmatrix}^T \begin{bmatrix} EA & 0 \\ 0 & EI \end{bmatrix} \begin{Bmatrix} \mathbf{N}'_u + \frac{\mathbf{N}_v}{R(s)} \\ \mathbf{N}'_\theta \end{Bmatrix} dx \begin{Bmatrix} u_1 \\ v_1 \\ \theta_1 \\ u_2 \\ v_2 \\ \theta_2 \end{Bmatrix} \quad (3.157)$$

where

$$\begin{aligned} \mathbf{N}_u &= [N_{uu1} \quad N_{uv1} \quad N_{u\theta 1} \quad N_{uu2} \quad N_{uv2} \quad N_{u\theta 2}] \\ \mathbf{N}_v &= [N_{vu1} \quad N_{vv1} \quad N_{v\theta 1} \quad N_{vu2} \quad N_{vv2} \quad N_{v\theta 2}] \\ \mathbf{N}_\theta &= [N_{\theta u1} \quad N_{\theta v1} \quad N_{\theta\theta 1} \quad N_{\theta u2} \quad N_{\theta v2} \quad N_{\theta\theta 2}] \end{aligned}$$

By substituting (3.156), the kinetic energy equation can be expressed in matrix forms as

$$K_E = \begin{Bmatrix} \ddot{u}_1 \\ \ddot{v}_1 \\ \ddot{\theta}_1 \\ \ddot{u}_2 \\ \ddot{v}_2 \\ \ddot{\theta}_2 \end{Bmatrix}^T \frac{1}{2}\int_0^L \begin{Bmatrix} \mathbf{N}_u \\ \mathbf{N}_v \\ \mathbf{N}_\theta \end{Bmatrix}^T \begin{bmatrix} \rho A & 0 & 0 \\ 0 & \rho A & 0 \\ 0 & 0 & \rho I \end{bmatrix} \begin{Bmatrix} \mathbf{N}_u \\ \mathbf{N}_v \\ \mathbf{N}_\theta \end{Bmatrix} dx \begin{Bmatrix} \ddot{u}_1 \\ \ddot{v}_1 \\ \ddot{\theta}_1 \\ \ddot{u}_2 \\ \ddot{v}_2 \\ \ddot{\theta}_2 \end{Bmatrix} \quad (3.158)$$

By substituting (3.156) into (3.155), the external work together with the boundary conditions, equation in matrix form becomes

$$W_E = \begin{Bmatrix} u_1 \\ v_1 \\ \theta_1 \\ u_2 \\ v_2 \\ \theta_2 \end{Bmatrix}^T \begin{Bmatrix} N_{uu1}P_1 \\ N_{vv1}Q_1 \\ N_{\theta\theta 1}M_1 \\ N_{uu2}P_2 \\ N_{vv2}Q_2 \\ N_{\theta\theta 2}M_2 \end{Bmatrix} - \begin{Bmatrix} u_1 \\ v_1 \\ \theta_1 \\ u_2 \\ v_2 \\ \theta_2 \end{Bmatrix}^T \int_0^L \begin{Bmatrix} N_{uu1}p \\ N_{vv1}q \\ N_{\theta\theta 1}m \\ N_{uu2}p \\ N_{vv2}q \\ N_{\theta\theta 2}m \end{Bmatrix} dx \quad (3.159)$$

The theory of minimum potential energy can be used conveniently to develop the finite element formulation for a general curved Euler-Bernoulli beam to obtain the equilibrium equation in matrix form.

$$\begin{aligned} \Pi &= \frac{1}{2}\mathbf{d}^T\mathbf{K}\mathbf{d} - \frac{1}{2}\ddot{\mathbf{d}}^T\mathbf{M}\ddot{\mathbf{d}} - \mathbf{d}^T\mathbf{f} \\ \frac{\partial \Pi}{\partial \mathbf{d}^T} &= \mathbf{K}\mathbf{d} - \mathbf{M}\ddot{\mathbf{d}} - \mathbf{f} = \mathbf{0} \end{aligned} \quad (3.160)$$

The stiffness matrix, mass matrix, and loading vector, are given by

$$\mathbf{K} = \int_0^L \begin{Bmatrix} \mathbf{N}'_u + \frac{\mathbf{N}_v}{R(s)} \\ \mathbf{N}'_\theta \end{Bmatrix}^T \begin{bmatrix} EA & 0 \\ 0 & EI \end{bmatrix} \begin{Bmatrix} \mathbf{N}'_u + \dfrac{\mathbf{N}_v}{R(s)} \\ \mathbf{N}'_\theta \end{Bmatrix} dx \tag{3.161}$$

$$\mathbf{M} = \int_0^L \begin{Bmatrix} \mathbf{N}_u \\ \mathbf{N}_v \\ \mathbf{N}_\theta \end{Bmatrix}^T \begin{bmatrix} \rho A & 0 & 0 \\ 0 & \rho A & 0 \\ 0 & 0 & \rho I \end{bmatrix} \begin{Bmatrix} \mathbf{N}_u \\ \mathbf{N}_v \\ \mathbf{N}_\theta \end{Bmatrix} dx \tag{3.162}$$

$$\mathbf{f} = \begin{Bmatrix} N_{uu1}P_1 \\ N_{vv1}Q_1 \\ N_{\theta\theta1}M_1 \\ N_{uu2}P_2 \\ N_{vv2}Q_2 \\ N_{\theta\theta2}M_2 \end{Bmatrix} + \int_0^L \begin{Bmatrix} N_{uu1}p \\ N_{vv1}q \\ N_{\theta\theta1}m \\ N_{uu2}p \\ N_{vv2}q \\ N_{\theta\theta2}m \end{Bmatrix} dx \tag{3.163}$$

3.9 Matrix Formulation of General Curved Timoshenko Beam

Following the same procedure with the previous section, in this section, the derivation of shape functions will not be presented due to the complexity of (3.104). Similarly, with the help of NURBS that will be discussed in the following chapter, the shape functions will be replaced by the NURBS functions that can be integrated numerically.

3.9.1 *Governing Equation in Matrix Forms*

The governing equation in the matrix forms can be obtained from the Eq. (3.85).

$$\delta H = \int_{t_1}^{t_2} (\delta S_E - \delta K_E - \delta W_E) dt = 0 \tag{3.164}$$

At an instant time of $t_1 = t_2$, the internal strain and kinetic energy of the beam are given as

$$S_E = \frac{1}{2}\int_0^L \left(\int_A \left\{ \begin{array}{l} \left(\frac{du}{ds} - r\frac{d\theta}{dx} + \frac{v}{R(s)}\right) E \left(\frac{du}{ds} - r\frac{d\theta}{dx} + \frac{v}{R(s)}\right) + \\ \left(\frac{dv}{ds} - \theta + \frac{u}{R(s)}\right) G \left(\frac{dv}{ds} - \theta + \frac{u}{R(s)}\right) \end{array} \right\} dA \right) ds \quad (3.165)$$

$$\begin{aligned} K_E &= \frac{1}{2}\int_0^L \left(\int_A \left(\frac{du}{dt} - y\frac{d\theta}{dt}\right) \rho \left(\frac{du}{dt} - y\frac{d\theta}{dt}\right) dA \right) ds \\ &+ \frac{1}{2}\int_0^L \left(\int_A \left(\frac{dv}{dt}\right) \rho \left(\frac{dv}{dt}\right) dA \right) ds \end{aligned} \quad (3.166)$$

with the external work defined by

$$W_E = \int_0^L \{p\ u + q\ v + m\ \theta\} dx \quad (3.167)$$

The derivatives of displacements and rotations can be represented by using the nodal general displacement vector $\mathbf{d} = \{\, u_1 \quad v_1 \quad \theta_1 \quad u_2 \quad v_2 \quad \theta_2 \,\}^T$ as given by

$$\begin{aligned} \frac{du}{ds} &= \left[\; N'_{uu1} \quad N'_{uv1} \quad N'_{u\theta 1} \quad N'_{uu2} \quad N'_{uv2} \quad N'_{u\theta 2} \;\right] \mathbf{d} \\ \frac{dv}{ds} &= \left[\; N'_{vu1} \quad N'_{vv1} \quad N'_{v\theta 1} \quad N'_{vu2} \quad N'_{vv2} \quad N'_{v\theta 2} \;\right] \mathbf{d} \\ \frac{d\theta}{ds} &= \left[\; N'_{\theta u1} \quad N'_{\theta v1} \quad N'_{\theta\theta 1} \quad N'_{\theta u2} \quad N'_{\theta v2} \quad N'_{\theta\theta 2} \;\right] \mathbf{d} \end{aligned} \quad (3.168)$$

By substituting (3.168), the internal strain energy equation can be expressed in matrix forms as

$$S_E = \frac{1}{2}\begin{Bmatrix} u_1 \\ v_1 \\ \theta_1 \\ u_2 \\ v_2 \\ \theta_2 \end{Bmatrix}^T \int_0^L \left\{ \begin{array}{l} \begin{Bmatrix} \mathbf{N}'_u + \frac{\mathbf{N}_v}{R(s)} \\ -\frac{\mathbf{N}_u}{R(s)} + \mathbf{N}'_v - \mathbf{N}_\theta \\ -\mathbf{N}'_\theta \end{Bmatrix}^T \begin{bmatrix} EA & 0 & 0 \\ 0 & \kappa GA & 0 \\ 0 & 0 & EI \end{bmatrix} \\ \times \begin{Bmatrix} \mathbf{N}'_u + \dfrac{\mathbf{N}_v}{R(s)} \\ -\dfrac{\mathbf{N}_u}{R(s)} + \mathbf{N}'_v - \mathbf{N}_\theta \\ -\mathbf{N}'_\theta \end{Bmatrix} \end{array} \right\} dx \begin{Bmatrix} u_1 \\ v_1 \\ \theta_1 \\ u_2 \\ v_2 \\ \theta_2 \end{Bmatrix} \quad (3.169)$$

where

$$\begin{aligned}
\mathbf{N}_u &= \begin{bmatrix} N_{uu1} & N_{uv1} & N_{u\theta1} & N_{uu2} & N_{uv2} & N_{u\theta2} \end{bmatrix} \\
\mathbf{N}_v &= \begin{bmatrix} N_{vu1} & N_{vv1} & N_{v\theta1} & N_{vu2} & N_{vv2} & N_{v\theta2} \end{bmatrix} \\
\mathbf{N}_\theta &= \begin{bmatrix} N_{\theta u1} & N_{\theta v1} & N_{\theta\theta1} & N_{\theta u2} & N_{\theta v2} & N_{\theta\theta2} \end{bmatrix}
\end{aligned}$$

By substituting (3.168), the kinetic energy equation can be expressed in matrix forms as

$$K_E = \begin{Bmatrix} \ddot{u}_1 \\ \ddot{v}_1 \\ \ddot{\theta}_1 \\ \ddot{u}_2 \\ \ddot{v}_2 \\ \ddot{\theta}_2 \end{Bmatrix}^T \frac{1}{2}\int_0^L \begin{Bmatrix} \mathbf{N}_u \\ \mathbf{N}_v \\ \mathbf{N}_\theta \end{Bmatrix}^T \begin{bmatrix} \rho A & 0 & 0 \\ 0 & \rho A & 0 \\ 0 & 0 & \rho I \end{bmatrix} \begin{Bmatrix} \mathbf{N}_u \\ \mathbf{N}_v \\ \mathbf{N}_\theta \end{Bmatrix} dx \begin{Bmatrix} \ddot{u}_1 \\ \ddot{v}_1 \\ \ddot{\theta}_1 \\ \ddot{u}_2 \\ \ddot{v}_2 \\ \ddot{\theta}_2 \end{Bmatrix} \tag{3.170}$$

By substituting (3.168) into (3.167), the external work together with the boundary conditions, equation in matrix form becomes

$$W_E = \begin{Bmatrix} u_1 \\ v_1 \\ \theta_1 \\ u_2 \\ v_2 \\ \theta_2 \end{Bmatrix}^T \begin{Bmatrix} N_{uu1}P_1 \\ N_{vv1}Q_1 \\ N_{\theta\theta1}M_1 \\ N_{uu2}P_2 \\ N_{vv2}Q_2 \\ N_{\theta\theta2}M_2 \end{Bmatrix} - \begin{Bmatrix} u_1 \\ v_1 \\ \theta_1 \\ u_2 \\ v_2 \\ \theta_2 \end{Bmatrix}^T \int_0^L \begin{Bmatrix} N_{uu1}p \\ N_{vv1}q \\ N_{\theta\theta1}m \\ N_{uu2}p \\ N_{vv2}q \\ N_{\theta\theta2}m \end{Bmatrix} dx \tag{3.171}$$

The theory of minimum potential energy can be used conveniently to develop the finite element formulation for a general curved Timoshenko beam to obtain the equilibrium equation in matrix form.

$$\begin{aligned}
\Pi &= \frac{1}{2}\mathbf{d}^T\mathbf{K}\mathbf{d} - \frac{1}{2}\ddot{\mathbf{d}}^T\mathbf{M}\ddot{\mathbf{d}} - \mathbf{d}^T\mathbf{f} \\
\frac{\partial \Pi}{\partial \mathbf{d}^T} &= \mathbf{K}\mathbf{d} - \mathbf{M}\ddot{\mathbf{d}} - \mathbf{f} = \mathbf{0}
\end{aligned} \tag{3.172}$$

The stiffness matrix, mass matrix, and loading vector are given by

$$\mathbf{K} = \int_0^L \left\{ \left\{ \begin{matrix} \mathbf{N}'_u + \frac{\mathbf{N}_v}{R(s)} \\ -\frac{\mathbf{N}_u}{R(s)} + \mathbf{N}'_v - \mathbf{N}_\theta \\ -\mathbf{N}'_\theta \end{matrix} \right\}^T \begin{bmatrix} EA & 0 & 0 \\ 0 & \kappa GA & 0 \\ 0 & 0 & EI \end{bmatrix} \times \left\{ \begin{matrix} \mathbf{N}'_u + \dfrac{\mathbf{N}_v}{R(s)} \\ -\dfrac{\mathbf{N}_u}{R(s)} + \mathbf{N}'_v - \mathbf{N}_\theta \\ -\mathbf{N}'_\theta \end{matrix} \right\} \right\} dx \tag{3.173}$$

$$\mathbf{M} = \int_0^L \left\{ \begin{matrix} \mathbf{N}_u \\ \mathbf{N}_v \\ \mathbf{N}_\theta \end{matrix} \right\}^T \begin{bmatrix} \rho A & 0 & 0 \\ 0 & \rho A & 0 \\ 0 & 0 & \rho I \end{bmatrix} \left\{ \begin{matrix} \mathbf{N}_u \\ \mathbf{N}_v \\ \mathbf{N}_\theta \end{matrix} \right\} dx \tag{3.174}$$

$$\mathbf{f} = \left\{ \begin{matrix} N_{uu1}P_1 \\ N_{vv1}Q_1 \\ N_{\theta\theta 1}M_1 \\ N_{uu2}P_2 \\ N_{vv2}Q_2 \\ N_{\theta\theta 2}M_2 \end{matrix} \right\} + \int_0^L \left\{ \begin{matrix} N_{uu1}p \\ N_{vv1}q \\ N_{\theta\theta 1}m \\ N_{uu2}p \\ N_{vv2}q \\ N_{\theta\theta 2}m \end{matrix} \right\} dx \tag{3.175}$$

3.10 Matrix Formulation of a Circular Curved Beam

The following sections will discuss how to obtain circular curved beam element formulations in the matrix and vector forms. The circular curved beam element formulation is a particular case of general curved beam element where the curvature of every point on the beam axis has the same radius of curvature from a single center of the circle.

For comparison purposes, the shape functions of the circular curved beam element were adopted from the paper of Litewka and Rakowski (1997, 1998). Hence, the parameters and notations from the paper are kept unchanged. Figure 3.15 shows a two-node circular curved beam element on the plane with six degrees of freedom and the radius of curvature R. The angle of the sector, length, and angular coordinate of the element are $\alpha, a, x(-\alpha_0 \leq x \leq \alpha_0)$, respectively.

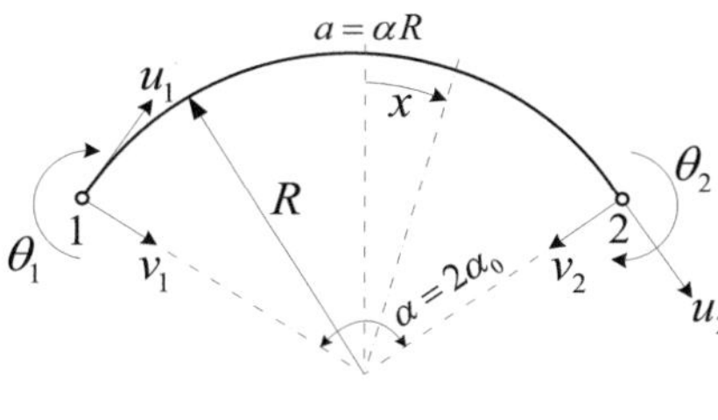

Fig. 3.15 Degrees of freedom of a circular curved beam element

3.10.1 Exact Shape Functions

The algebraic-trigonometric functions of the angular coordinate s are given by

$$\begin{aligned}\mathbf{N}_u(x) &= \mathbf{C}_{u0} + \mathbf{C}_{u1}x + (\mathbf{C}_{u2} + \mathbf{C}_{u3}x)\sin(x) + (\mathbf{C}_{u4} + \mathbf{C}_{u5}x)\cos(x)\\ \mathbf{N}_v(x) &= \mathbf{C}_{v0} + \mathbf{C}_{v1}x + (\mathbf{C}_{v2} + \mathbf{C}_{v3}x)\sin(x) + (\mathbf{C}_{v4} + \mathbf{C}_{v5}x)\cos(x)\\ \mathbf{N}_\theta(x) &= \mathbf{C}_{\theta 0} + \mathbf{C}_{\theta 1}x + (\mathbf{C}_{\theta 2} + \mathbf{C}_{\theta 3}x)\sin(x) + (\mathbf{C}_{\theta 4} + \mathbf{C}_{\theta 5}x)\cos(x)\end{aligned} \tag{3.176}$$

To obtain all the coefficients in the shape functions, unit support displacements were imposed to the clamped-clamped circular curved beam element, and internal forces are determined using the flexibility method. Then, using the principle of virtual work, the displacement functions of u, v, θ along the arc by taking into account the influences of bending, shear, and axial forces were solved. From there, a full set of 18 shape functions related to the unit support displacements are derived. The lists of shape functions $\mathbf{N}_u, \mathbf{N}_v, \mathbf{N}_\theta$ are shown in Figs. 3.16 and 3.17. The figures of the shape functions $\mathbf{N}_u, \mathbf{N}_v, \mathbf{N}_\theta$ are shown in Figs. 3.18, 3.19, and 3.20 for different values of $\varphi = 0$ (Euler-Bernoulli beam assumption), 0.1, and 1.0 to show the effects of the shear deformation. The formulas are using the following parameters and notations for simplicity. It can be observed that by setting $d = 0$, the shear deformation effect will be constrained. Thus, the rotation of beam cross section will follow the Euler-Bernoulli assumptions.

The following notations are adopted for simplicity:
$d = \frac{EI}{\kappa GA} \cdot \frac{1}{a^2}$; $d_1 = \alpha^2 d$; $e = \frac{EI}{EA} \cdot \frac{1}{a^2}$; $e_1 = \alpha^2 e$; $s = \sin\alpha$; $s_0 = \sin\alpha_0$.
$c = \cos\alpha$; $c_0 = \cos\alpha_0$; $A' = 1 + d_1 + e_1$; $A'' = 1 - d_1 + e_1$.
$D_1 = \alpha(\alpha + \sin\alpha)A' - 2[2(1 - \cos\alpha) + \alpha d_1 \sin\alpha]$; $D' = 1/D_1 + 1/D_2$.
$D_2 = \alpha(\alpha + \sin\alpha)A' - 2\alpha\sin\alpha(1 + e_1)$; $D'' = 1/D_1 - 1/D_2$.

The coefficients $\mathbf{C}_{u/v/\theta\text{-}0}$~$\mathbf{C}_{u/v/\theta\text{-}5}$ in (3.176) for radial displacement, tangential displacement, and rotation shape functions of a circular curved beam element are shown explicitly in Figs. 3.16 and 3.17.

The stiffness matrix and loading vector are defined by

$$\mathbf{K} = \int_0^L \left\{ \left\{ \begin{array}{c} \mathbf{N}_u' + \frac{\mathbf{N}_v}{R} \\ -\frac{\mathbf{N}_u}{R} + \mathbf{N}_v' - \mathbf{N}_\theta \\ \mathbf{N}_\theta' \end{array} \right\}^T \begin{bmatrix} EA & 0 & 0 \\ 0 & \kappa GA & 0 \\ 0 & 0 & EI \end{bmatrix} \times \left\{ \begin{array}{c} \mathbf{N}_u' + \dfrac{\mathbf{N}_v}{R} \\ -\dfrac{\mathbf{N}_u}{R} + \mathbf{N}_v' - \mathbf{N}_\theta \\ \mathbf{N}_\theta' \end{array} \right\} \right\} dx \tag{3.177}$$

$$
\begin{aligned}
N_{uu1} &= -\frac{1}{D_2}\alpha s + \left(-\frac{2}{D_1}s\right)x + \left\{-\frac{1}{2}\left[D'c + D''\right]A''\alpha c_0 + \frac{2}{D_1}(\alpha + s)c_0 - \frac{A'}{2D_2}\alpha^2 s_0\right\}\sin x + \frac{A'}{D_2}\alpha s_0\ x\sin x \\
&\quad + \left\{\frac{1}{2}\left[D'c + D''\right]A''\alpha s_0 + \frac{2}{D_2}\alpha s_0 + \frac{A'}{2D_1}\alpha^2 c_0 - \frac{2}{D_1}\alpha s_0 s\right\}\cos x - \frac{A'}{D_1}\alpha c_0\ x\cos x
\end{aligned}
$$

$$
\begin{aligned}
N_{uu2} &= -\frac{1}{D_2}\alpha s + \frac{2}{D_1}s\ x + \left\{\frac{1}{2}\left[D' + D''c\right]A''\alpha c_0 - \frac{2}{D_1}(\alpha + s)c_0 - \frac{A'}{2D_2}\alpha^2 s_0 + s_0\right\}\sin x + \frac{A'}{D_2}\alpha s_0\ x\sin x \\
&\quad + \left\{-\frac{1}{2}\left[D' + D''c\right]A''\alpha s_0 + \frac{2}{D_2}\alpha s_0 - \frac{A'}{2D_1}\alpha^2 c_0 + \frac{2}{D_1}s_0 s + c_0\right\}\cos x + \frac{A'}{D_1}\alpha c_0\ x\cos x
\end{aligned}
$$

$$
\begin{aligned}
N_{uv1} &= \frac{1}{D_2}\alpha(1+c) - \frac{2}{D_1}(1-c)x + \left\{-\frac{1}{2}D'sA''\alpha c_0 + \frac{A'}{2D_2}\alpha^2 c_0 + \frac{2}{D_1}\left(\alpha s_0 + (1-c)c_0\right)\right\}\sin x \\
&\quad - \frac{A'}{D_2}\alpha c_0\ x\sin x + \left\{\frac{1}{2}D'sA''\alpha s_0 + \frac{A'}{2D_1}\alpha^2 s_0 - 2\left[\frac{1}{D_1}(1-c)s_0 + \frac{1}{D_2}\alpha c_0\right]\right\}\cos x - \frac{A'}{D_1}\alpha s_0\ x\cos x
\end{aligned}
$$

$$
\begin{aligned}
N_{uv2} &= -\frac{1}{D_2}\alpha(1+c) - \frac{2}{D_1}(1-c)x + \left\{-\frac{1}{2}D''sA''\alpha c_0 - \frac{A'}{2D_2}\alpha^2 c_0 + \frac{2}{D_1}\left(\alpha s_0 + (1-c)c_0\right) + c_0\right\}\sin x \\
&\quad + \frac{A'}{D_2}\alpha c_0\ x\sin x + \left\{\frac{1}{2}D''sA''\alpha s_0 + \frac{A'}{2D_1}\alpha^2 s_0 + 2\left[-\frac{1}{D_1}(1-c)s_0 + \frac{1}{D_2}\alpha c_0\right] - s_0\right\}\cos x - \frac{A'}{D_1}\alpha s_0\ x\cos x
\end{aligned}
$$

$$
\begin{aligned}
N_{u\theta 1} &= \left(\frac{1}{D_2}s + \frac{1}{2\alpha}\right) + \frac{2}{\alpha^2}\left(\frac{\alpha s}{D_1} - \frac{1}{2} - \frac{2(1-c)}{D_1}\right)x + \left\{\begin{array}{l} \frac{1}{2}\left[D'c + D'' - \frac{2s}{\alpha D_1}\right]A''c_0 + \frac{A'}{2D_2}\alpha s_0 \\ + \frac{2}{D_1}\left(\frac{2}{\alpha}s_0 - c_0\right) + \frac{2c_0}{\alpha^2 D_1}[2(1-c) - \alpha s] + \frac{c_0}{\alpha^2} \end{array}\right\}\sin x \\
&\quad - \frac{A'}{D_2}s_0\ x\sin x - \left\{\begin{array}{l} \frac{1}{2}\left[D'c + D'' - \frac{2s}{\alpha D_1}\right]A''s_0 + \frac{2}{D_2}s_0 \\ + \frac{A'}{2D_1}(\alpha c_0 - s_0) + \frac{2s_0}{\alpha^2 D_1}[2(1-c) - \alpha s] + \frac{s_0}{\alpha^2} \end{array}\right\}\cos x + \frac{A'}{\alpha D_1}(\alpha c_0 - 2s_0)\ x\cos x
\end{aligned}
$$

$$
\begin{aligned}
N_{u\theta 2} &= \left(\frac{1}{D_2}s + \frac{1}{2\alpha}\right) - \frac{2}{\alpha^2}\left(\frac{\alpha s}{D_1} - \frac{1}{2} - \frac{2(1-c)}{D_1}\right)x + \left\{\begin{array}{l} -\frac{1}{2}\left[D' + D''c - \frac{2s}{\alpha D_1}\right]A''c_0 + \frac{A'}{2D_2}\alpha s_0 - \frac{s_0}{\alpha} \\ + \frac{2}{D_1}\left(-\frac{2}{\alpha}s_0 + c_0\right) + \frac{2c_0}{\alpha^2 D_1}[2(1-c) - \alpha s] - \frac{c_0}{\alpha^2} \end{array}\right\}\sin x \\
&\quad - \frac{A'}{D_2}s_0\ x\sin x + \left\{\begin{array}{l} \frac{1}{2}\left[D' + D''c - \frac{2s}{\alpha D_1}\right]A''s_0 - \frac{2}{D_2}s_0 - \frac{c_0}{\alpha} \\ + \frac{A'}{2D_1}(\alpha c_0 - 2s_0) + \frac{2s_0}{\alpha^2 D_1}[2(1-c) - \alpha s] + \frac{s_0}{\alpha^2} \end{array}\right\}\cos x - \frac{A'}{\alpha D_1}(\alpha c_0 - 2s_0)\ x\cos x
\end{aligned}
$$

Fig. 3.16 Circular curved beam tangential nodal displacement shape function: N_u

$$N_{vu1}=-\frac{2}{D_1}s+\left\{-\frac{A'}{2D_1}\alpha^2 c_0-\frac{1}{2}D'A''c_0\alpha s+\frac{2}{D_1}s_0 s\right\}\sin x+\frac{A'}{D_1}\alpha c_0\ x\sin x$$
$$+\left\{-\frac{A'}{2D_2}\alpha^2 s_0+\frac{1}{2}D'A''s_0\alpha s+\frac{2}{D_1}c_0 s\right\}\cos x+\frac{A'}{D_2}\alpha s_0\ x\cos x$$

$$N_{vu2}=\frac{2}{D_1}s+\left\{\frac{A'}{2D_1}\alpha^2 c_0+\frac{1}{2}D''A''c_0\alpha s-\frac{2}{D_1}s_0 s-c_0\right\}\sin x-\frac{A'}{D_1}\alpha c_0\ x\sin x$$
$$+\left\{-\frac{A'}{2D_2}\alpha^2 s_0-\frac{1}{2}D''A''s_0\alpha s-\frac{2}{D_1}c_0 s+s_0\right\}\cos x+\frac{A'}{D_2}\alpha s_0\ x\cos x$$

$$N_{vv1}=-\frac{2}{D_1}(1-c)+\left\{\frac{1}{2}[D'c-D'']A''\alpha c_0-\frac{A'}{2D_1}\alpha^2 s_0+\frac{2}{D_1}(1-c)s_0\right\}\sin x$$
$$+\frac{A'}{D_1}\alpha s_0\ x\sin x+\left\{-\frac{1}{2}[D'c-D'']A''\alpha s_0+\frac{A'}{2D_2}\alpha^2 c_0+\frac{2}{D_1}(1-c)c_0\right\}\cos x-\frac{A'}{D_2}\alpha c_0\ x\cos x$$

$$N_{vv2}=-\frac{2}{D_1}(1-c)+\left\{-\frac{1}{2}[D'-D''c]A''\alpha c_0+s_0-\frac{A'}{2D_1}\alpha^2 s_0+\frac{2}{D_1}(1-c)s_0\right\}\sin x$$
$$+\frac{A'}{D_1}\alpha s_0\ x\sin x+\left\{\frac{1}{2}[D'-D''c]A''\alpha s_0+c_0-\frac{A'}{2D_2}\alpha^2 c_0+\frac{2}{D_1}(1-c)c_0\right\}\cos x+\frac{A'}{D_2}\alpha c_0\ x\cos x$$

$$N_{v\theta1}=-\frac{1}{\alpha^2}\left(\frac{2}{D_1}[2(1-c)-\alpha s]+1\right)+\left\{\begin{array}{l}\frac{1}{2}\left[D's-\frac{2}{\alpha D_1}(1-c)\right]A''c_0+\frac{A'}{D_1}\left(\frac{1}{2}\alpha c_0-s_0\right)\\+\frac{2s_0}{\alpha^2 D_1}[2(1-c)-\alpha s]+\frac{s_0}{\alpha^2}\end{array}\right\}\sin x$$
$$-\frac{A'}{\alpha D_1}(\alpha c_0-2s_0)\,x\sin x-\left\{\begin{array}{l}\frac{1}{2}\left[D's-\frac{2}{\alpha D_1}(1-c)\right]A''s_0+\frac{A'}{2D_2}\alpha s_0\\+\frac{2c_0}{\alpha^2 D_1}[2(1-c)-\alpha s]+\frac{c_0}{\alpha^2}\end{array}\right\}\cos x-\frac{A'}{D_2}s_0\ x\cos x$$

$$N_{v\theta2}=\frac{1}{\alpha^2}\left(\frac{2}{D_1}[2(1-c)-\alpha s]+1\right)+\left\{\begin{array}{l}-\frac{1}{2}\left[D''s-\frac{2}{\alpha D_1}(1-c)\right]A''c_0-\frac{A'}{D_1}\left(\frac{1}{2}\alpha c_0-s_0\right)\\+\frac{c_0}{\alpha}-\frac{2s_0}{\alpha^2 D_1}[2(1-c)-\alpha s]-\frac{s_0}{\alpha^2}\end{array}\right\}\sin x$$
$$+\frac{A'}{\alpha D_1}(\alpha c_0-2s_0)\,x\sin x+\left\{\begin{array}{l}\frac{1}{2}\left[D''s-\frac{2}{\alpha D_1}(1-c)\right]A''s_0+\frac{A'}{2D_2}\alpha s_0-\frac{s_0}{\alpha}\\-\frac{2c_0}{\alpha^2 D_1}[2(1-c)-\alpha s]-\frac{c_0}{\alpha^2}\end{array}\right\}\cos x-\frac{A'}{D_2}s_0\ x\cos x$$

$$N_{\theta u1}=-\frac{1}{RD_2}\alpha cs+\left(-\frac{2s}{RD_1}\right)x+\left(\frac{2\alpha c_0}{RD_1}\right)\sin x+\left(\frac{2\alpha s_0}{RD_2}\right)\cos x$$

$$N_{\theta u2}=-\frac{1}{RD_2}\alpha cs+\left(\frac{2s}{RD_1}\right)x+\left(-\frac{2\alpha c_0}{RD_1}\right)\sin x+\left(\frac{2\alpha s_0}{RD_2}\right)\cos x$$

$$N_{\theta v1}=\frac{\alpha(1+c)}{RD_2}+\left(-\frac{2(1-c)}{RD_1}\right)x+\left(\frac{2\alpha s_0}{RD_1}\right)\sin x+\left(-\frac{2\alpha c_0}{RD_2}\right)\cos x$$

$$N_{\theta v1}=\frac{\alpha(1+c)}{RD_2}+\left(-\frac{2(1-c)}{RD_1}\right)x+\left(\frac{2\alpha s_0}{RD_1}\right)\sin x+\left(-\frac{2\alpha c_0}{RD_2}\right)\cos x$$

$$N_{\theta\theta1}=\left(\frac{1}{D_2}\alpha s+\frac{1}{2}\right)-\frac{1}{\alpha}\left(\frac{2}{D_1}[2(1-c)-\alpha s]+1\right)x+\frac{2}{D_1}(2s_0-\alpha c_0)\sin x+\left(-\frac{2\alpha s_0}{D_2}\right)\cos x$$

$$N_{\theta\theta2}=\left(\frac{1}{D_2}\alpha s+\frac{1}{2}\right)+\frac{1}{\alpha}\left(\frac{2}{D_1}[2(1-c)-\alpha s]+1\right)x-\frac{2}{D_1}(2s_0-\alpha c_0)\sin x+\left(-\frac{2\alpha s_0}{D_2}\right)\cos x$$

Fig. 3.17 Circular curved beam radial displacement and rotational shape functions: N_v, N_θ

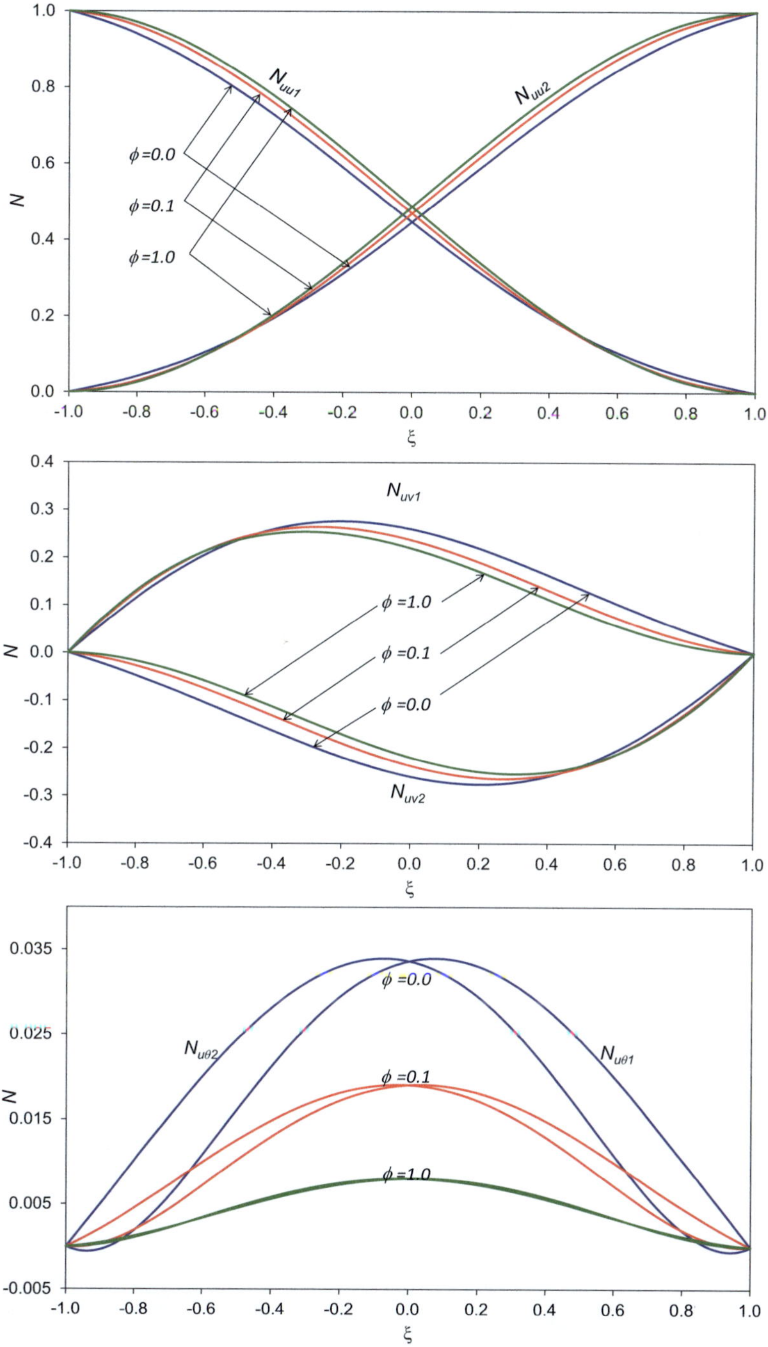

Fig. 3.18 Circular curved beam tangential displacement shape function: N_u with $\varphi = 0$, 0.1 and 1.0

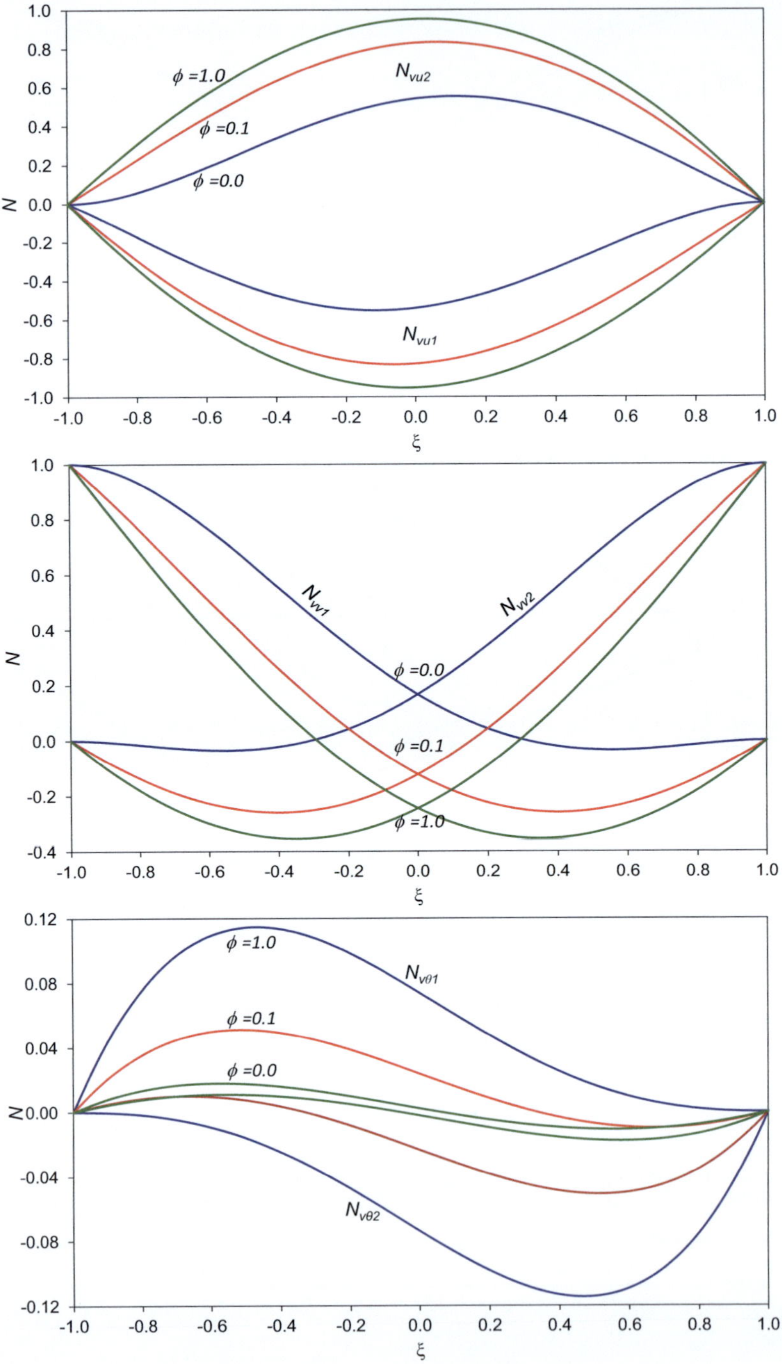

Fig. 3.19 Circular curved beam radial displacement shape function: N_v with $\varphi = 0$, 0.1 and 1.0

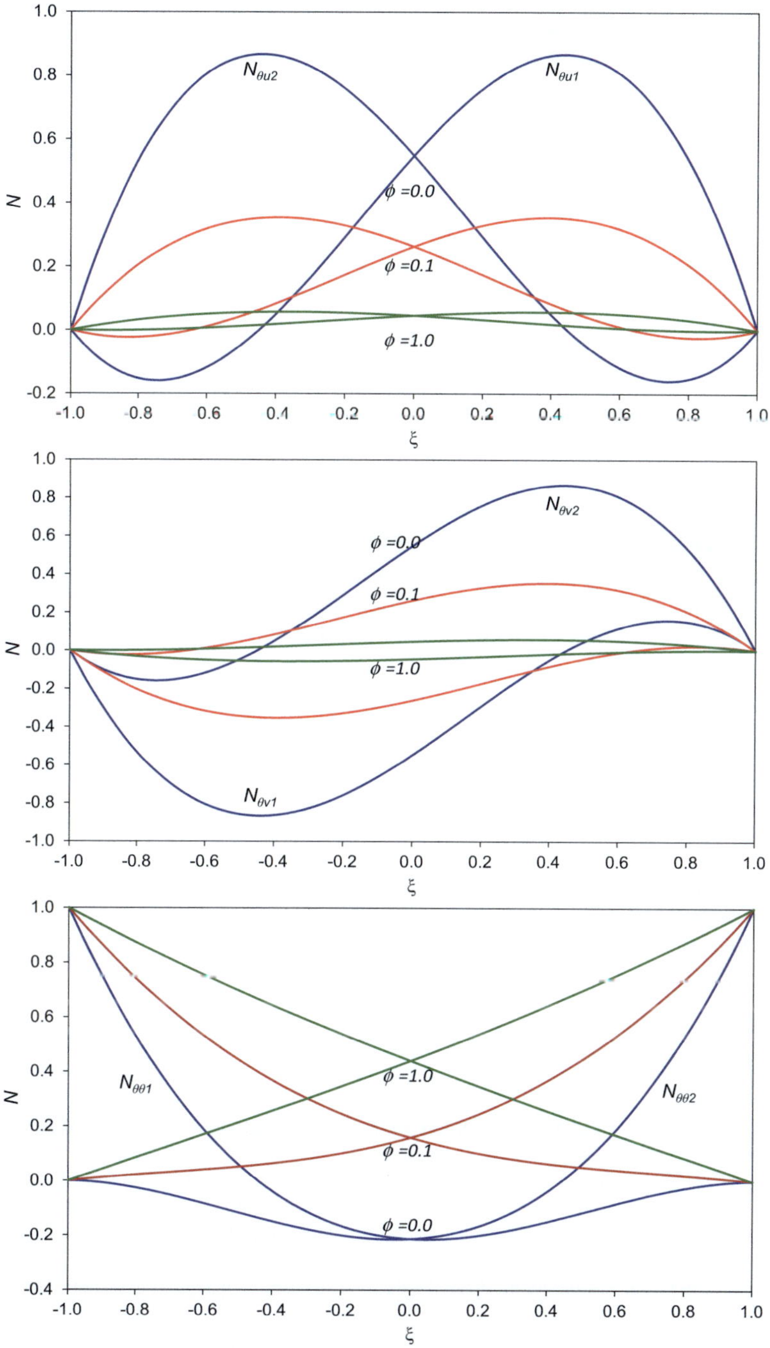

Fig. 3.20 Circular curved beam rotational shape function: N_θ with $\varphi = 0$, 0.1 and 1.0

$$\mathbf{M} = \int_0^L \begin{Bmatrix} \mathbf{N}_u \\ \mathbf{N}_v \\ \mathbf{N}_\theta \end{Bmatrix}^T \begin{bmatrix} \rho A & 0 & 0 \\ 0 & \rho A & 0 \\ 0 & 0 & \rho I \end{bmatrix} \begin{Bmatrix} \mathbf{N}_u \\ \mathbf{N}_v \\ \mathbf{N}_\theta \end{Bmatrix} dx \qquad (3.178)$$

$$\mathbf{f} = \begin{Bmatrix} N_{uu1}P_1 \\ N_{vv1}Q_1 \\ N_{\theta\theta1}M_1 \\ N_{uu2}P_2 \\ N_{vv2}Q_2 \\ N_{\theta\theta2}M_2 \end{Bmatrix} + \int_0^L \begin{Bmatrix} N_{uu1}p \\ N_{vv1}q \\ N_{\theta\theta1}m \\ N_{uu2}p \\ N_{vv2}q \\ N_{\theta\theta2}m \end{Bmatrix} dx \qquad (3.179)$$

It can be seen that the shape functions described in (3.176) are too complex to be manipulated in the stiffness and mass matrices formulation by hand, even to type into a computer program. The other option to determine the matrices is to integrate numerically, however, due to a high order of degree polynomials that must be squared, and hence many Gauss points are required for computing the matrices. Therefore, computationally it becomes inefficient.

We can show later in the following chapter that by using the NURBS functions, the general curved beam element formulation becomes a very easy and straightforward method to solve any kind of curved-type beam elements, including the circular curved beam type.

References

Litewka P, Rakowski J (1997) An efficient curved beam finite element. Int J Num Meth Engrg 40:2629–2652

Litewka P, Rakowski J (1998) The exact thick arch finite element. Comput Struct 68:369–379

Further Reading

Bathe KJ (1982) Finite element procedures in engineering analysis. Prentice-Hall, Englewood Cliffs

Cook RD (1981) Concepts and applications of finite element analysis, 2nd edn. Wiley, New York

Fung YC (1969) A first course in continuum mechanics. Prentice-Hall, Englewood Cliffs

Shames IH, Dym CL (1985) Energy and finite element methods in structural mechanics. Hemispere Publishing, New York

Washizu K (1982) Variational methods in elasticity & plasticity, 3rd edn. Pergamon Press, Oxford

Chapter 4
Isogeometric Approach to Beam Element

4.1 Introduction

The isogeometric approach is pioneered by several researchers (Basilevs et al. 2010; Cottrell et al. 2006, 2007, 2009; Gontier and Vollmer 1995; Schramm and Pilkey 1993), where the NURBS curves are used to represent the real geometry and direct mesh modeling in the finite element context. The NURBS with its informative geometrical information which is usually created in the CAD environment can be transported seamlessly into the finite element formulation. In the formulation of the NURBS-based element, the shape function which is a fundamental component in element formulation can also be constructed straightforwardly.

Table 4.1 shows the matching relationship between the NURBS and finite element concepts. This comparison will help the reader who is familiar with the finite element concept to understand the isogeometric approach.

4.2 NURBS for Straight Beam Element Formulation

4.2.1 Euler-Bernoulli Beam Element

A straight Euler-Bernoulli beam element is considered as a one-dimensional problem. In the isogeometric approach, the coordinates of the geometry of the beam can be represented by using the NURBS curve (Sect. 1.3.8) as follows:

$$x = \sum_{i=0}^{n} S_{i,p}(\xi)\ P_{xi} \qquad -1 \le \xi \le 1 \tag{4.1}$$

B.S. Gan, *An Isogeometric Approach to Beam Structures*,
DOI 10.1007/978-3-319-56493-7_4

Table 4.1 Interpretation of the isogeometric approach in the beam finite element concept

Isogeometric analysis	Finite element analysis
Exact geometry	Approximate geometry
Control points	Nodal points
Control variables	Nodal variables
NURBS shape function	Polynomial shape function
Changeable level of C^{i}-continuity	C^{0}-continuity

where

$$S_{i,p}(\xi) = \frac{N_{i,p}(\xi)\ w_i}{\sum_{j=0}^{n} N_{j,p}(\xi)\ w_j} \qquad w_i > 0$$

In isogeometric approach, the classical shape functions, $\mathbf{N}_{u/v/\theta}(\xi)$, and their derivatives of the Euler-Bernoulli straight beam element in (3.119) are substituted by the rational basis functions for the B-spline curve, $S_{i,p}(\xi)$, from (1.18) and the derivative of the rational basis function $S_{i,p}^{k}(\xi)$ which can be computed from (1.20).

The displacements and rotation of any arbitrary point along the straight Euler-Bernoulli beam as shown in (3.114) can be calculated from the general nodal displacement of the beam element by using the shape functions given as

$$u = \sum_{i=0}^{nu} S_{i,pu}(\xi)\ u_i\,; \qquad v = \sum_{i=0}^{nv} S_{i,pv}(\xi)\ v_i\,; \qquad \theta = \frac{dv}{d\xi} = \sum_{i=0}^{nv} S_{i,pv}^{1}(\xi)\ \theta_i \tag{4.2}$$

It is worth to be noted that, in isogeometric approach, the nodal displacements and rotations are treated independently by using each rational basis B-spline functions. Thus they are not coupled in the formulation.

Depending on the order of the *p-th* degree polynomial and knot vector (Ξ) being selected, we need to prepare the corresponding number of control points (*CP*) required. The increasing of control points being used to construct the NURBS curve will increase the degree-of-freedom of the beam as illustrated in Fig. 4.1. The figure shows the illustration of degree-of-freedom at each control points at the beam only when the general displacements have the same degree of polynomial.

The equilibrium equation is obtained by minimizing the potential energy, given as

$$\begin{aligned} \Pi &= \frac{1}{2}\mathbf{d}^T\mathbf{K}\mathbf{d} - \frac{1}{2}\ddot{\mathbf{d}}^T\mathbf{M}\ddot{\mathbf{d}} - \mathbf{d}^T\mathbf{f} \\ \frac{\partial \Pi}{\partial \mathbf{d}^T} &= \mathbf{K}\mathbf{d} - \mathbf{M}\ddot{\mathbf{d}} - \mathbf{f} = \mathbf{0} \end{aligned} \tag{4.3}$$

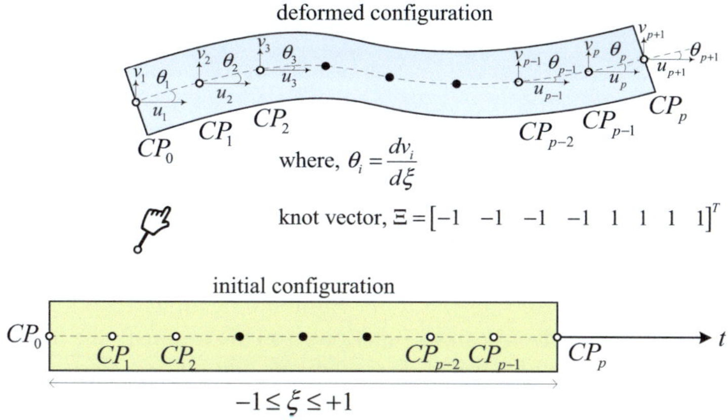

Fig. 4.1 Conceptual kinematic of Euler-Bernoulli beam element in isogeometric approach

The classical Euler-Bernoulli beam shape functions were incomplete polynomials due to the assumption of a cross section of the beam which remains perpendicular to the beam axis, where the derivative of vertical displacement assumption was used for the slope of the beam. However, in the isogeometric approach using the NURBS, a complete series of rational basis B-spline functions are necessary to compose a function of the order *p-th* degree curve.

Therefore, the series of rational basis B-spline functions of the NURBS curve cannot be used directly to compose the shape functions of the Euler-Bernoulli beam. To apply the rational basis B-spline functions for the vertical displacement and rotational for the Euler-Bernoulli beam shape functions, selective basis functions $B_{i,pv}(\xi_j)$ are taken from the series of rational basis B-spline functions of the NURBS that will be customized and used to fit the classical shape functions.

Referring to (3.124), the stiffness matrix is given as

$$\mathbf{K} = \begin{bmatrix} \mathbf{k}_{11} & 0 \\ 0 & \mathbf{k}_{22} \end{bmatrix}$$

$$\mathbf{k}_{11} = \sum_{j=1}^{ngu} \left[\frac{EA}{J_j^2} \left(\sum_{i=0}^{nu} S_{i,pu}^1(\xi_j) \right)^T \left(\sum_{i=0}^{nu} S_{i,pu}^1(\xi_j) \right) \right] Det(J_j)\ w_j \tag{4.4}$$

$$\mathbf{k}_{22} = \sum_{j=1}^{ngv} \left[\frac{EI}{J_j^4} \left(\sum_{i=0}^{nv} B_{i,pv}^2(\xi_j) \right)^T \left(\sum_{i=0}^{nv} B_{i,pv}^2(\xi_j) \right) \right] Det(J_j)\ w_j$$

In (4.4), the sub-matrix $\mathbf{k}_{22}$ has a dimension of 4x4 which includes the vertical displacement and rotational stiffness of the beam. The selective basis functions are given as

$$\begin{aligned}
B_{0,pv}(\xi_j) &= S_{0,pv}(\xi_j) + S_{1,pv}(\xi_j)\\
B_{1,pv}(\xi_j) &= S_{2,pv}(\xi_j) + S_{3,pv}(\xi_j)\\
B_{2,pv}(\xi_j) &= \frac{L}{3} S_{1,pv}(\xi_j)\\
B_{3,pv}(\xi_j) &= -\frac{L}{3} S_{2,pv}(\xi_j)
\end{aligned} \tag{4.5}$$

It was found that the complete NURBS basis functions to represent a polynomial with order p cannot be used for the Euler-Bernoulli beam assumption. The determination of the above selective basis functions is made to satisfy the four unknown boundary conditions of two vertical displacements and two rotations of the beam.

Referring to (3.125), the mass matrix is given by

$$\begin{aligned}
\mathbf{M} &= \begin{bmatrix} \mathbf{m}_{11} & 0 \\ 0 & \mathbf{m}_{22} \end{bmatrix}\\
\mathbf{m}_{11} &= \sum_{j=1}^{ngu} \left[\rho A \left(\sum_{i=0}^{nu} S_{i,pu}(\xi_j) \right)^T \left(\sum_{i=0}^{nu} S_{i,pu}(\xi_j) \right) \right] Det(J_j)\ w_j\\
\mathbf{m}_{22} &= \sum_{j=1}^{ngv} \left[\rho A \left(\sum_{i=0}^{nv} B_{i,pv}(\xi_j) \right)^T \left(\sum_{i=0}^{nv} B_{i,pv}(\xi_j) \right) \right] Det(J_j)\ w_j
\end{aligned} \tag{4.6}$$

The displacements and acceleration vectors are given as

$$\begin{aligned}
\mathbf{u} &= [u_i \ \ . \ . \ . \ \ u_{i=pu+1}]^T \quad ; \quad \ddot{\mathbf{u}} = [\ddot{u}_i \ \ . \ . \ . \ \ \ddot{u}_{i=pu+1}]^T\\
\mathbf{v} &= [v_i \ \ . \ . \ . \ \ v_{i=pv+1}]^T \quad ; \quad \ddot{\mathbf{v}} = [\ddot{v}_i \ \ . \ . \ . \ \ \ddot{v}_{i=pv+1}]^T\\
\boldsymbol{\theta} &= [\theta_i \ \ . \ . \ . \ \ \theta_{i=pv+1}]^T \quad ; \quad \ddot{\boldsymbol{\theta}} = [\ddot{\theta}_i \ \ . \ . \ . \ \ \ddot{\theta}_{i=pv+1}]^T
\end{aligned} \tag{4.7}$$

Referring to (3.126), the loading vector is given by

$$\mathbf{f} = \begin{Bmatrix} \mathbf{P} \\ \mathbf{Q} \\ \mathbf{M} \end{Bmatrix} + \begin{Bmatrix} \sum_{j=1}^{ngu} \left[\left(\sum_{i=0}^{nu} S_{i,pu}(\xi_j) \right) p \right] Det(J_j)\ w_j \\ \sum_{j=1}^{ngv} \left[\left(\sum_{i=0}^{(nv+1)/2-1} B_{i,pu}(\xi_j) \right) q \right] Det(J_j)\ w_j \\ \sum_{j=1}^{ngv} \left[\left(\sum_{i=(nv+1)/2}^{nv} B_{i,pu}(\xi_j) \right) m \right] Det(J_j)\ w_j \end{Bmatrix} \tag{4.8}$$

4.2.2 *KMmatrixEBNURBS Program List*

MATLAB code KMmatrixEBNURBS.m and functions: Legendre.m (Sect. 2.3.2) and DNurbsLeibnitz.m (Sect. 1.3.14) integrate the stiffness and mass matrices of the straight Euler-Bernoulli beam example (Sect. 3.6.3) numerically by using NURBS functions. The computed matrices of beam example in Fig. 4.2 are shown in Fig. 4.3.

```
% KMmatrixEBNURBS.m
% Construct matrices K and M
% Based on the Euler-Bernoulli Beam Theory using NURBS shape functions
clear all;clc;
L  = 1.3;   E = 1525; A  = 0.2;
I  = 0.014; rho = 630;
```

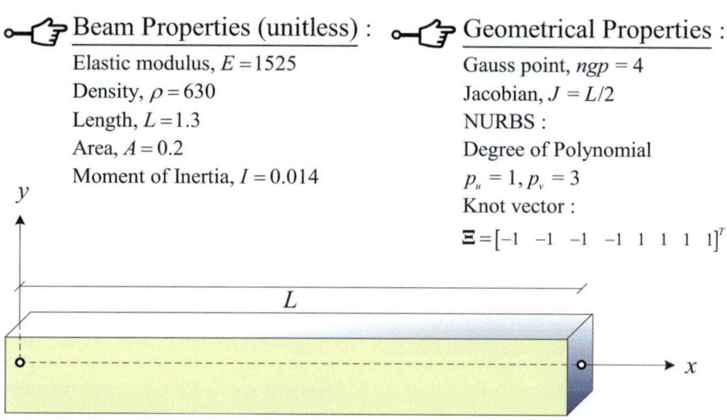

Fig. 4.2 Straight Euler-Bernoulli beam example

The Stiffness Matrix

234.6154	0	0	-234.6154	0	0
0	116.6136	75.7988	0	-116.6136	75.7988
0	75.7988	65.6923	0	-75.7988	32.8462
-234.6154	0	0	234.6154	0	0
0	-116.6136	-75.7988	0	116.6136	-75.7988
0	75.7988	32.8462	0	-75.7988	65.6923

The Mass Matrix

54.6000	0	0	27.3000	0	0
0	68.9815	12.0360	0	12.9185	-5.7090
0	12.0360	4.1652	0	5.7090	-2.3595
27.3000	0	0	54.6000	0	0
0	12.9185	5.7090	0	68.9815	-12.0360
0	-5.7090	-2.3595	0	-12.0360	4.1652

Fig. 4.3 The stiffness and mass matrices of a straight Euler-Bernoulli beam element example

```
% NURBS for u displacement data
npa  = 1;
knota = [-ones(1,npa+1) ones(1,npa+1)]; knota = [knota ones(1,npa)];
nba  = size(knota,2)-npa-1;
wta  = ones(nba-npa);
% NURBS for v and theta displacement data
npb  = 3;
knotb = [-ones(1,npb+1) ones(1,npb+1)]; knotb = [knotb ones(1,npb)];
nbb  = size(knotb,2)-npb-1;
wtb  = ones(nbb-npb);
% Jacobi and Gauss
Jac  = L/2;
ngp = 4;  Gpw = Legendre(ngp);
dim = npa+1+npb+1;          Bmtx = zeros(1,4);
Kmtx = zeros(dim,dim);      Mmtx = zeros(dim,dim);
Su0  = zeros(nba-npa,ngp);  Su1  = zeros(nba-npa,ngp);
Sv0  = zeros(nbb-npb,ngp);
St0  = zeros(nbb-npb,ngp);  Sv2  = zeros(nbb-npb,ngp);
for i=1:ngp
 t = Gpw(i,1); % coordinate
 DNurbsbasis = DNurbsLeibnitz(npa,knota,wta,t);
 Su0(1:nba-npa,i)=DNurbsbasis(1:nba-npa,npa+5); % non-derivative
 Su1(1:nba-npa,i)=DNurbsbasis(1:nba-npa,npa+7); % 1st-derivative
end
for i=1:ngp
 t = Gpw(i,1); % coordinate
 DNurbsbasis = DNurbsLeibnitz(npb,knotb,wtb,t);
 Sv0(1:nbb-npb,i)=DNurbsbasis(1:nbb-npb,npb+5); % non-derivative
 Sv1(1:nbb-npb,i)=DNurbsbasis(1:nbb-npb,npb+7); % 1st-derivative
 Sv2(1:nbb-npb,i)=DNurbsbasis(1:nbb-npb,npb+8); % 2nd-derivative
end
% K11
for i=1:ngp
 for j=1:nba-npa
  for k=1:nba-npa
      Kmtx(j,k) = Kmtx(j,k)+E*A*Su1(j,i)/Jac*Su1(k,i)/Jac*Jac*Gpw
      (i,2);
  end
 end
end
% K22
nna = nba-npa; nnb = nbb-npb;
for i=1:ngp
 Bmtx(1) = Sv2(1,i)/Jac/Jac + Sv2(2,i)/Jac/Jac;
 Bmtx(2) = Sv2(3,i)/Jac/Jac + Sv2(4,i)/Jac/Jac;
```

```
 Bmtx(3) = L/3*Sv2(2,i)/Jac/Jac;
 Bmtx(4) =-L/3*Sv2(3,i)/Jac/Jac;
 for j=1:4
   for k=1:4
      Kmtx(j+nna,k+nna) = Kmtx(j+nna,k+nna)...
    +E*I*Bmtx(j)*Bmtx(k)*Jac*Gpw(i,2);
   end
 end
end
% Swapping the matrices to the general displacement vector
Kmtx=Kmtx([1,3,5,2,4,6],:); % row
Kmtx=Kmtx(:,[1,3,5,2,4,6]); % column
disp('The Stiffness Matrix');
disp(Kmtx);
% Mass matrix K
% M11 translation mass rho*A
for i=1:ngp
 for j=1:nna
  for k=1:nna
   Mmtx(j,k) = Mmtx(j,k)+rho*A*Su0(j,i)*Su0(k,i)*Jac*Gpw(i,2);
  end
 end
end
% M22A translation mass rho*A
for i=1:ngp
 Bmtx(1) = Sv0(1,i) + Sv0(2,i);
 Bmtx(2) = Sv0(3,i) + Sv0(4,i);
 Bmtx(3) = L/3*Sv0(2,i);
 Bmtx(4) =-L/3*Sv0(3,i);
 for j=1:4
  for k=1:4
   Mmtx(j+nna,k+nna) = Mmtx(j+nna,k+nna)...
    +rho*A*Bmtx(j)*Bmtx(k)*Jac*Gpw(i,2);
  end
 end
end
% M22B rotatory mass rho*I
for i=1:ngp
 Bmtx(1) = Sv1(1,i)/Jac + Sv1(2,i)/Jac;
 Bmtx(2) = Sv1(3,i)/Jac + Sv1(4,i)/Jac;
 Bmtx(3) = L/3*Sv1(2,i)/Jac;
 Bmtx(4) =-L/3*Sv1(3,i)/Jac;
 for j=1:4
  for k=1:4
```

```
   Mmtx(j+nna,k+nna) = Mmtx(j+nna,k+nna)...
    +rho*I*Bmtx(j)*Bmtx(k)*Jac*Gpw(i,2);
  end
 end
end
% Swapping the matrices to the general displacement vector
Mmtx=Mmtx([1,3,5,2,4,6],:); % row
Mmtx=Mmtx(:,[1,3,5,2,4,6]); % column
disp('The Mass Matrix');
disp(Mmtx);
```

We can see that the values of computed stiffness and mass matrices shown in Fig. 3.11 and Fig. 4.3 are exactly the same. The stiffness and mass matrices values from both polynomial functions and NURBS functions can be verified by using the classical formulations (see Appendix A).

4.2.3 Timoshenko Beam Element

A straight Timoshenko beam element is considered as a one-dimensional problem. In the isogeometric approach, the coordinates of the geometry of the beam can be represented by using the NURBS curve (Sect. 1.3.8) as follows:

$$x = \sum_{i=0}^{n} S_{i,p}(\xi)\ P_{xi} \qquad -1 \leq \xi \leq 1 \tag{4.9}$$

where

$$S_{i,p}(\xi) = \frac{N_{i,p}(\xi)\ w_i}{\sum_{j=0}^{n} N_{j,p}(\xi)\ w_j} \qquad w_i > 0$$

In the isogeometric approach, the classical shape functions, $\mathbf{N}_{u/v/\theta}(\xi)$, and their derivatives of the Timoshenko straight beam element in (3.136), (3.138), and (3.139) are substituted by using the rational B-spline basis functions of the NURBS curve. The rational basis function $S_{i,p}(\xi)$ is given in (1.18), and the derivative of the rational basis function $S^k_{i,p}(\xi)$ can be computed from (1.20).

The displacements and rotation of any arbitrary point along the straight Timoshenko beam can be calculated from the general nodal displacement of the beam element by using the rational B-spline basis functions which are given by

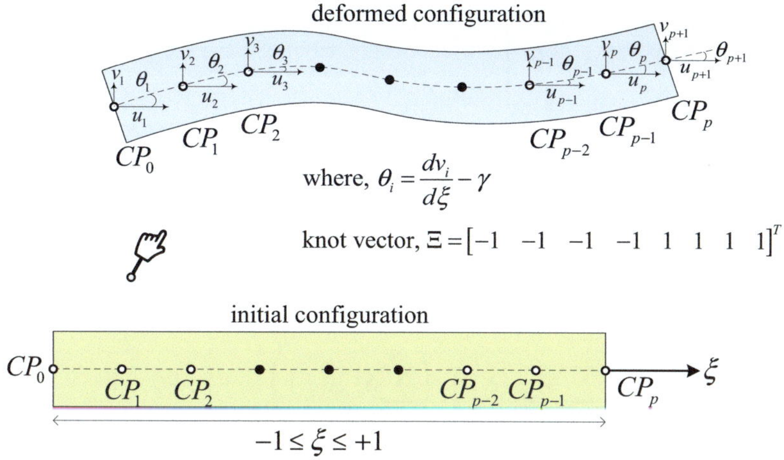

Fig. 4.4 Conceptual kinematic of Timoshenko beam element in isogeometric approach

$$u = \sum_{i=0}^{nu} S_{i,pu}(\xi)\, u_i\,; \qquad v = \sum_{i=0}^{nv} S_{i,pv}(\xi)\, v_i\,; \qquad \theta = \sum_{i=0}^{n\theta} S_{i,p\theta}(\xi)\, \theta_i \tag{4.10}$$

Depending on the order of the *p-th* degree polynomial and knot vector (Ξ) being selected, we need to prepare the corresponding number of control points (*CP*) required. The increasing of control points being used to construct the NURBS curve will increase the degree-of-freedom of the beam as illustrated in Fig. 4.4. The figure shows the illustration of degree-of-freedom at each control points at the beam only when the general displacements have the same degree of polynomial.

The equilibrium equation is obtained by minimizing the potential energy, given as

$$\begin{aligned} \Pi &= \frac{1}{2}\mathbf{d}^T\mathbf{K}\mathbf{d} - \frac{1}{2}\ddot{\mathbf{d}}^T\mathbf{M}\ddot{\mathbf{d}} - \mathbf{d}^T\mathbf{f} \\ \frac{\partial \Pi}{\partial \mathbf{d}^T} &= \mathbf{K}\mathbf{d} - \mathbf{M}\ddot{\mathbf{d}} - \mathbf{f} = \mathbf{0} \end{aligned} \tag{4.11}$$

Referring to (3.149), the stiffness matrix is given as

$$
\mathbf{K} = \begin{bmatrix} \mathbf{k}_{11} & 0 & 0 \\ 0 & \mathbf{k}_{22} & 0 \\ 0 & 0 & \mathbf{k}_{33} \end{bmatrix}
$$

$$
\begin{aligned}
\mathbf{k}_{11} &= \sum_{j=1}^{ngu} \left[\frac{EA}{J_j^2} \left(\sum_{i=0}^{nu} S_{i,pu}^1(\xi_j) \right)^T \left(\sum_{i=0}^{nu} S_{i,pu}^1(\xi_j) \right) \right] Det(J_j) \ w_j \\
\mathbf{k}_{22} &= \sum_{j=1}^{ngv} \left[\frac{\kappa GA}{J_j^2} \left(\sum_{i=0}^{nv} S_{i,pv}^1(\xi_j) \right)^T \left(\sum_{i=0}^{nv} S_{i,pv}^1(\xi_j) \right) \right] Det(J_j) \ w_j \\
\mathbf{k}_{33} &= \sum_{j=1}^{ng\theta} \left[\begin{array}{l} \frac{EI}{J_j^2} \left(\sum_{i=0}^{n\theta} S_{i,p\theta}^1(\xi_j) \right)^T \left(\sum_{i=0}^{n\theta} S_{i,p\theta}^1(\xi_j) \right) \\ -\kappa GA \left(\sum_{i=0}^{nv} S_{i,p\theta}(\xi_j) \right)^T \left(\sum_{i=0}^{nv} S_{i,p\theta}(\xi_j) \right) \end{array} \right] Det(J_j) \ w_j
\end{aligned} \tag{4.12}
$$

The displacements and acceleration vectors are given as

$$
\begin{aligned}
\mathbf{u} &= [u_i \ \ . \ \ . \ \ . \ \ u_{i=pu+1}]^T \quad ; \quad \ddot{\mathbf{u}} = [\ddot{u}_i \ \ . \ \ . \ \ . \ \ \ddot{u}_{i=pu+1}]^T \\
\mathbf{v} &= [v_i \ \ . \ \ . \ \ . \ \ v_{i=pv+1}]^T \quad ; \quad \ddot{\mathbf{v}} = [\ddot{v}_i \ \ . \ \ . \ \ . \ \ \ddot{v}_{i=pv+1}]^T \\
\boldsymbol{\theta} &= [\theta_i \ \ . \ \ . \ \ . \ \ \theta_{i=p\theta+1}]^T \quad ; \quad \ddot{\boldsymbol{\theta}} = [\ddot{\theta}_i \ \ . \ \ . \ \ . \ \ \ddot{\theta}_{i=p\theta+1}]^T
\end{aligned} \tag{4.13}
$$

Referring to (3.150), the mass matrix is given by

$$
\mathbf{M} = \begin{bmatrix} \mathbf{m}_{11} & 0 & 0 \\ 0 & \mathbf{m}_{22} & 0 \\ 0 & 0 & \mathbf{m}_{33} \end{bmatrix}
$$

$$
\begin{aligned}
\mathbf{m}_{11} &= \sum_{j=1}^{ngu} \left[\rho A \left(\sum_{i=0}^{nu} S_{i,pu}(\xi_j) \right)^T \left(\sum_{i=0}^{nu} S_{i,pu}(\xi_j) \right) \right] Det(J_j) \ w_j \\
\mathbf{m}_{22} &= \sum_{j=1}^{ngv} \left[\rho A \left(\sum_{i=0}^{nv} S_{i,pv}(\xi_j) \right)^T \left(\sum_{i=0}^{nv} S_{i,pv}(\xi_j) \right) \right] Det(J_j) \ w_j \\
\mathbf{m}_{33} &= \sum_{j=1}^{ng\theta} \left[\rho I \left(\sum_{i=0}^{n\theta} S_{i,p\theta}(\xi_j) \right)^T \left(\sum_{i=0}^{n\theta} S_{i,p\theta}(\xi_j) \right) \right] Det(J_j) \ w_j
\end{aligned} \tag{4.14}
$$

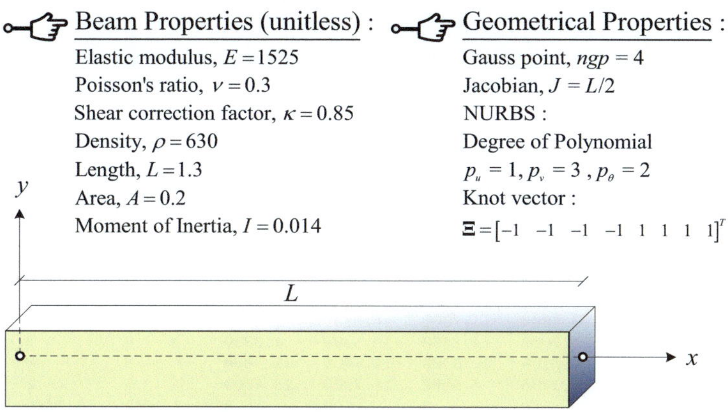

Fig. 4.5 Straight Timoshenko beam example

Referring to (3.151), the loading vector is given by

$$\mathbf{f} = \begin{Bmatrix} \mathbf{P} \\ \mathbf{Q} \\ \mathbf{M} \end{Bmatrix} + \begin{Bmatrix} \sum_{j=1}^{ngu} \left[\left(\sum_{i=0}^{nu} S_{i,pu}(\xi_j) \right) p \right] Det(J_j) \ \ w_j \\ \sum_{j=1}^{ngv} \left[\left(\sum_{i=0}^{nv} S_{i,pv}(\xi_j) \right) q \right] Det(J_j) \ \ w_j \\ \sum_{j=1}^{ng\theta} \left[\left(\sum_{i=0}^{n\theta} S_{i,p\theta}(\xi_j) \right) m \right] Det(J_j) \ \ w_j \end{Bmatrix} \tag{4.15}$$

4.2.4 KMmatrixTiNURBS Program List

MATLAB code KMmatrixTiNURBS.m and functions: Legendre.m (Sect. 2.3.2) and DNurbsLeibnitz.m (Sect. 1.3.14) integrate numerically the stiffness and mass matrices of a straight Timoshenko beam example (Sect. 3.7.3) by using NURBS functions. The computed matrices of beam example in Fig. 4.5 are shown in Fig. 4.6.

```
% KMmatrixTiNURBS.m
% Construct matrices K and M
% Based on the Timoshenko Beam Theory by using NURBS shape functions
clear all;clc;
L  = 1.3;  E = 1525; A = 0.2; nu = 0.3;
I  = 0.014; kp = 0.85; rho = 630;
G  = E/2/(1+nu);
% NURBS for u displacement data
npa  = 1;
```

The Stiffness Matrix

234.6154	-234.6154	0	0	0	0	0	0	0
-234.6154	234.6154	0	0	0	0	0	0	0
0	0	138.0621	-69.0311	-46.0207	-23.0104	59.8269	29.9135	9.9712
0	0	-69.0311	92.0414	23.0104	-46.0207	-29.9135	9.9712	19.9423
0	0	-46.0207	23.0104	92.0414	-69.0311	-19.9423	-9.9712	29.9135
0	0	-23.0104	-46.0207	-69.0311	138.0621	-9.9712	-29.9135	-59.8269
0	0	59.8269	-29.9135	-19.9423	-9.9712	47.8224	2.0138	-6.6279
0	0	29.9135	9.9712	-9.9712	-29.9135	2.0138	39.1808	2.0138
0	0	9.9712	19.9423	29.9135	-59.8269	-6.6279	2.0138	47.8224

The Mass Matrix

54.6000	27.3000	0	0	0	0	0	0	0
27.3000	54.6000	0	0	0	0	0	0	0
0	0	23.4000	11.7000	4.6800	1.1700	0	0	0
0	0	11.7000	14.0400	10.5300	4.6800	0	0	0
0	0	4.6800	10.5300	14.0400	11.7000	0	0	0
0	0	1.1700	4.6800	11.7000	23.4000	0	0	0
0	0	0	0	0	0	2.2932	1.1466	0.3822
0	0	0	0	0	0	1.1466	1.5288	1.1466
0	0	0	0	0	0	0.3822	1.1466	2.2932

Fig. 4.6 The stiffness and mass matrices of a straight Timoshenko beam element example

```
knota = [-ones(1,npa+1) ones(1,npa+1)]; knota = [knota ones(1,npa)];
nba  = size(knota,2)-npa-1;
wta  = ones(nba-npa);
% NURBS for v displacement data
npb  = 3;
knotb = [-ones(1,npb+1) ones(1,npb+1)]; knotb = [knotb ones(1,npb)];
nbb  = size(knotb,2)-npb-1;
wtb  = ones(nbb-npb);
% NURBS for th displacement data
npc  = 2;
knotc = [-ones(1,npc+1) ones(1,npc+1)]; knotc = [knotc ones(1,npc)];
nbc  = size(knotc,2)-npc-1;
wtc  = ones(nbc-npc);
% Jacobi and Gauss
Jac  = L/2;
ngp = 4; Gpw = Legendre(ngp);
dim = npa+1+npb+1+npc+1;
Kmtx = zeros(dim,dim);   Mmtx = zeros(dim,dim);
Su0  = zeros(nba-npa,ngp); Sv0  = zeros(nbb-npb,ngp);
St0  = zeros(nbc-npc,ngp); Su1  = zeros(nba-npa,ngp);
Sv1  = zeros(nbb-npb,ngp); St1  = zeros(nbc-npc,ngp);
for i=1:ngp
 t = Gpw(i,1); % coordinate
 DNurbsbasis = DNurbsLeibnitz(npa,knota,wta,t);
 Su0(1:nba-npa,i)=DNurbsbasis(1:nba-npa,npa+5); % non-derivative
 Su1(1:nba-npa,i)=DNurbsbasis(1:nba-npa,npa+7); % 1st-derivative
end
```

```
for i=1:ngp
 t = Gpw(i,1); % coordinate
 DNurbsbasis = DNurbsLeibnitz(npb,knotb,wtb,t);
 Sv0(1:nbb-npb,i)=DNurbsbasis(1:nbb-npb,npb+5); % non-derivative
 Sv1(1:nbb-npb,i)=DNurbsbasis(1:nbb-npb,npb+7); % 1st-derivative
end
for i=1:ngp
 t = Gpw(i,1); % coordinate
 DNurbsbasis = DNurbsLeibnitz(npc,knotc,wtc,t);
 St0(1:nbc-npc,i)=DNurbsbasis(1:nbc-npc,npc+5); % non-derivative
 St1(1:nbc-npc,i)=DNurbsbasis(1:nbc-npc,npc+7); % 1st-derivative
end
% K11
for i=1:ngp
 for j=1:nba-npa
  for k=1:nba-npa
   Kmtx(j,k) = Kmtx(j,k)+E*A*Su1(j,i)/Jac*Su1(k,i)/Jac*Jac*Gpw(i,2);
  end
 end
end
% K22
nna = nba-npa; nnb = nbb-npb;
for i=1:ngp
 for j=1:nnb
  for k=1:nnb
   Kmtx(j+nna,k+nna) = Kmtx(j+nna,k+nna)...
     +kp*G*A*Sv1(j,i)/Jac*Sv1(k,i)/Jac*Jac*Gpw(i,2);
  end
 end
end
% K23
nnc = nbc-npc;
for i=1:ngp
 for j=1:nnb
  for k=1:nnc
   Kmtx(j+nna,k+nna+nnb) = Kmtx(j+nna,k+nna+nnb)...
     -kp*G*A*Sv1(j,i)/Jac*St0(k,i)*Jac*Gpw(i,2);
  end
 end
end
% K32
for i=1:ngp
 for j=1:nnc
  for k=1:nnb
```

```
   Kmtx(j+nna+nnb,k+nna) = Kmtx(j+nna+nnb,k+nna)...
     -kp*G*A*St0(j,i)*Sv1(k,i)/Jac*Jac*Gpw(i,2);
  end
 end
end
% K33
for i=1:ngp
 for j=1:nnc
  for k=1:nnc
   Kmtx(j+nna+nnb,k+nna+nnb) = Kmtx(j+nna+nnb,k+nna+nnb)...
     +kp*G*A*St0(j,i)*St0(k,i)*Jac*Gpw(i,2)...
     +E*I*St1(j,i)/Jac*St1(k,i)/Jac*Jac*Gpw(i,2);
  end
 end
end
disp('The Stiffness Matrix');
disp(Kmtx);
% Mass matrix K
% M11
for i=1:ngp
 for j=1:nna
  for k=1:nna
   Mmtx(j,k) = Mmtx(j,k)+rho*A*Su0(j,i)*Su0(k,i)*Jac*Gpw(i,2);
  end
 end
end
% M22
for i=1:ngp
 for j=1:nnb
  for k=1:nnb
   Mmtx(j+nna,k+nna) = Mmtx(j+nna,k+nna)...
     +rho*A*Sv0(j,i)*Sv0(k,i)*Jac*Gpw(i,2);
  end
 end
end
% M33
for i=1:ngp
 for j=1:nnc
  for k=1:nnc
   Mmtx(j+nna+nnb,k+nna+nnb) = Mmtx(j+nna+nnb,k+nna+nnb)...
     +rho*I*St0(j,i)*St0(k,i)*Jac*Gpw(i,2);
  end
 end
end
disp('The Mass Matrix');
disp(Mmtx);
```

We can see from Fig. 4.6 that the computed stiffness and mass matrices have nine degree-of-freedom beam element, since the horizontal displacement, u, v, and rotational θ were assumed to be linear, third order, and second order of NURBS, respectively. The stiffness and mass matrices of the element will be condensed and rearranged to suit the standard two-node beam element with six degree-of-freedoms in the following chapter.

4.3 NURBS for General Curved Beam Element Formulation

4.3.1 General Curved Euler-Bernoulli Beam

Even the geometry of general curved beam elements are constructed by the x and y coordinates on a plane, it can be considered as a one-dimensional problem. The path of the curve is measured as the local coordinate ξ. In the isogeometric approach, the coordinate of the geometry of the beam can be represented by using the NURBS curves (Sect. 1.3.8) as follow:

$$\begin{aligned} x &= \sum_{i=0}^{n} S_{i,p}(\xi)\ P_{xi} \\ y &= \sum_{i=0}^{n} S_{i,p}(\xi)\ P_{yi} \qquad\qquad -1 \le \xi \le 1 \end{aligned} \tag{4.16}$$

where

$$S_{i,p}(\xi) = \frac{N_{i,p}(\xi)\ w_i}{\sum_{j=0}^{n} N_{j,p}(\xi)\ w_j} \qquad\qquad w_i > 0$$

In isogeometric approach, the classical shape functions, $\mathbf{N}_{u/v/\theta}(\xi)$, and the derivatives of the general curved Euler-Bernoulli beam element in (3.156) are substituted by the rational basis functions from the B-spline curve $S_{i,p}(\xi)$ and the derivative of the rational basis function $S^k_{i,p}(\xi)$ which can be computed from (1.18).

The displacements and rotation of any arbitrary point along the general curved Euler-Bernoulli beam can be calculated from the general nodal displacement of the beam element by using the rational B-spline basis functions which are given by

$$u = \sum_{i=0}^{nu} S_{i,pu}(\xi)\ u_i\,; \qquad v = \sum_{i=0}^{nv} S_{i,pv}(\xi)\ v_i\,; \qquad \theta = \sum_{i=0}^{n\theta} S_{i,p\theta}(\xi)\ \theta_i \tag{4.17}$$

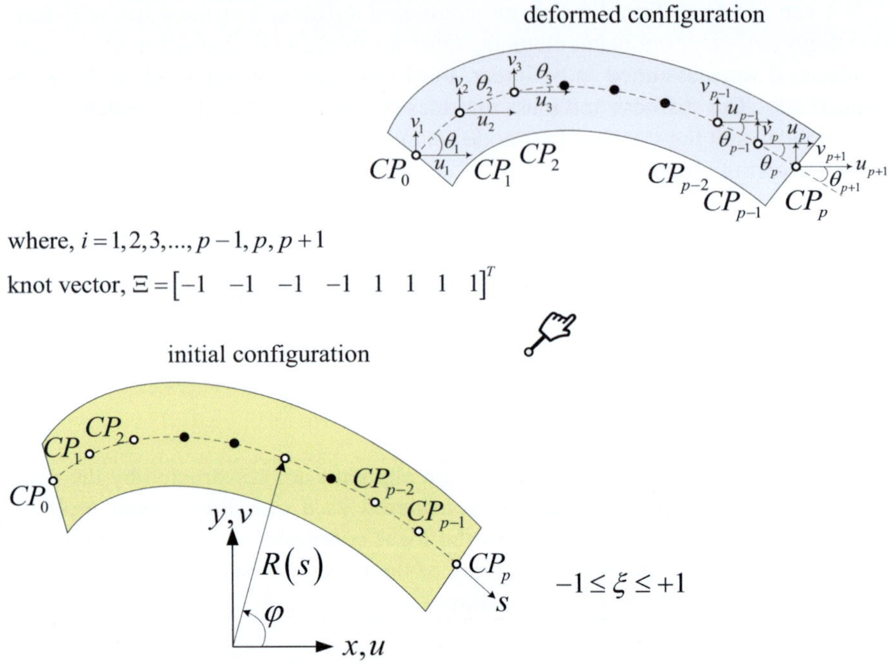

Fig. 4.7 Conceptual kinematic of general curved Euler-Bernoulli beam element in isogeometric approach

In this section, only the general curved Euler-Bernoulli beam element will be discussed. The general curved Euler-Bernoulli beam element can also be obtained by setting a large value of shear stiffness G in the general curved Timoshenko beam element formulation to model the cross sections that are not deformed under the shearing forces.

Depending on the order of the *p-th* degree polynomial and knot vector (Ξ) being selected, we need to prepare the corresponding number of control points (CP) required. The increasing of control points being used to construct the NURBS curve will increase the degree-of-freedom of the beam as illustrated in Fig. 4.7. The figure shows the illustration of degree-of-freedom at each control points at the beam only when the general displacements have the same degree of polynomial.

The equilibrium equation is obtained by minimizing the potential energy, given as

$$\begin{aligned} \Pi &= \frac{1}{2}\mathbf{d}^T\mathbf{K}\mathbf{d} - \frac{1}{2}\ddot{\mathbf{d}}^T\mathbf{M}\ddot{\mathbf{d}} - \mathbf{d}^T\mathbf{f} \\ \frac{\partial \Pi}{\partial \mathbf{d}^T} &= \mathbf{K}\mathbf{d} - \mathbf{M}\ddot{\mathbf{d}} - \mathbf{f} = \mathbf{0} \end{aligned} \tag{4.18}$$

Referring to (3.161), the stiffness matrix is given as

$$\mathbf{K} = [\mathbf{k}_{uv} + \mathbf{k}_{\theta}]$$

$$\mathbf{k}_{uv} = \sum_{j=1}^{ngu=ngv} \begin{bmatrix} \frac{EA}{J_j^2}\left(\sum_{k=0}^{nu} B^1_{k,pu}(\xi_i)\right)^T \left(\sum_{k=0}^{nu} B^1_{k,pu}(\xi_j)\right) \\ +\frac{EA}{J_j R(\xi_j)}\left(\sum_{k=0}^{nv} B_{k,pv}(\xi_i)\right)^T \left(\sum_{k=0}^{nv} B_{k,pv}(\xi_j)\right) \end{bmatrix} \times Det(J_j)\ w_j \tag{4.19}$$

$$\mathbf{k}_{\theta} = \sum_{j=1}^{ng\theta} \left[\frac{EI}{J_j^2}\left(\sum_{k=0}^{n\theta} B^1_{k,p\theta}(\xi_j)\right)^T \left(\sum_{k=0}^{n\theta} B^1_{k,p\theta}(\xi_j)\right)\right] \times Det(J_i)\ w_j$$

In (4.19), the derivatives of shape functions $B_{i,u/v/\theta}$ are not independent since they must fulfill the compatibility of no shear deformation condition of the beam cross section that includes the horizontal displacement, vertical displacement and rotational stiffness of the beam. Therefore, selective basis functions have to be determined from a fifth-order polynomial function with six unknowns that can be found from the boundary condition of a two-node 6 DOF beam element.

It was found that the complete NURBS basis functions to represent a polynomial with order p cannot be used for the general curved Euler-Bernoulli beam element. The determination of the above selective basis functions $\mathbf{B}$ has to satisfy the six unknown boundary conditions of two vertical displacements and two rotations of the beam.

The displacements and acceleration vectors are given as

$$\begin{aligned} \mathbf{u} &= [u_i \quad . \quad . \quad . \quad u_{i=pu+1}]^T \quad ; \quad \ddot{\mathbf{u}} = [\ddot{u}_i \quad . \quad . \quad . \quad \ddot{u}_{i=pu+1}]^T \\ \mathbf{v} &= [v_i \quad . \quad . \quad . \quad v_{i=pv+1}]^T \quad ; \quad \ddot{\mathbf{v}} = [\ddot{v}_i \quad . \quad . \quad . \quad \ddot{v}_{i=pv+1}]^T \\ \boldsymbol{\theta} &= [\theta_i \quad . \quad . \quad . \quad \theta_{i=p\theta+1}]^T \quad ; \quad \ddot{\boldsymbol{\theta}} = [\ddot{\theta}_i \quad . \quad . \quad . \quad \ddot{\theta}_{i=p\theta+1}]^T \end{aligned} \tag{4.20}$$

Referring to (3.162), the mass matrix is given by

$$\mathbf{M} = \begin{bmatrix} \mathbf{m}_{11} & 0 & 0 \\ 0 & \mathbf{m}_{22} & 0 \\ 0 & 0 & \mathbf{m}_{33} \end{bmatrix}$$

$$\begin{aligned} \mathbf{m}_{11} &= \sum_{j=1}^{ngu}\left[\rho A\left(\sum_{i=0}^{nu} B_{i,pu}(\xi_j)\right)^T\left(\sum_{i=0}^{nu} B_{i,pu}(\xi_j)\right)\right] Det(J_j)\ w_j \\ \mathbf{m}_{22} &= \sum_{j=1}^{ngv}\left[\rho A\left(\sum_{i=0}^{nv} B_{i,pv}(\xi_j)\right)^T\left(\sum_{i=0}^{nv} B_{i,pv}(\xi_j)\right)\right] Det(J_j)\ w_j \\ \mathbf{m}_{33} &= \sum_{j=1}^{ng\theta}\left[\rho I\left(\sum_{i=0}^{n\theta} B_{i,p\theta}(\xi_j)\right)^T\left(\sum_{i=0}^{n\theta} B_{i,p\theta}(\xi_j)\right)\right] Det(J_j)\ w_j \end{aligned} \tag{4.21}$$

Referring to (3.163), the loading vector is given by

$$\mathbf{f}=\begin{Bmatrix}\mathbf{P}\\ \mathbf{Q}\\ \mathbf{M}\end{Bmatrix}+\begin{Bmatrix}\sum_{j=1}^{ngu}\left[\left(\sum_{i=0}^{nu}B_{i,pu}(\xi_j)\right)p\right]Det(J_j)\ w_j\\ \sum_{j=1}^{ngv}\left[\left(\sum_{i=0}^{nv}B_{i,pv}(\xi_j)\right)q\right]Det(J_j)\ w_j\\ \sum_{j=1}^{ng\theta}\left[\left(\sum_{i=0}^{n\theta}B_{i,p\theta}(\xi_j)\right)m\right]Det(J_j)\ w_j\end{Bmatrix} \tag{4.22}$$

4.3.2 General Curved Timoshenko Beam

Similar with the general curved Euler-Bernoulli beam, the geometry of general curved Timoshenko beam elements are constructed by the x and y coordinates on a plane, it can be considered as a one-dimensional problem. The path of the curve is measured as the local coordinate ξ. In the isogeometric approach, the coordinate of the geometry of the beam can be represented by using the NURBS curves (Sect. 1.3.8) as follows:

$$\begin{aligned} x &= \sum_{i=0}^{n} S_{i,p}(\xi)\ P_{xi} \\ y &= \sum_{i=0}^{n} S_{i,p}(\xi)\ P_{yi} \qquad -1\le\xi\le 1 \end{aligned} \tag{4.23}$$

where

$$S_{i,p}(\xi)=\frac{N_{i,p}(\xi)\ w_i}{\sum_{j=0}^{n}N_{j,p}(\xi)\ w_j}\qquad w_i>0$$

In the isogeometric approach, the classical shape functions, $\mathbf{N}_{u/v/\theta}(\xi)$, and the derivatives of the Timoshenko straight beam element in (3.168) are substituted by the rational basis functions from the B-spline curve $S_{i,p}(\xi)$ and the derivative of the rational basis function $S^{k}_{i,p}(\xi)$ which can be computed from (1.18).

The displacements and rotation of any arbitrary point along the general curved Timoshenko beam can be calculated from the general nodal displacement of the beam element by using the rational B-spline basis functions which are given by

$$u = \sum_{i=0}^{nu} S_{i,pu}(\xi)\, u_i\,; \qquad v = \sum_{i=0}^{nv} S_{i,pv}(\xi)\, v_i\,; \qquad \theta = \sum_{i=0}^{n\theta} S_{i,p\theta}(\xi)\, \theta_i \tag{4.24}$$

It is worth to be noted that in the isogeometric approach, unlike in the case of Euler-Bernoulli beam element where the displacements and rotational are constrained by the no shear deformation assumption, for Timoshenko beam element, the nodal displacements and rotations are treated independently by using each rational basis B-spline functions. Thus they are not coupled in the formulation.

The general curved Euler-Bernoulli beam element can be done easily by setting a large value of shear stiffness G to model the cross sections that are not deformed under the shearing forces.

Depending on the order of the *p-th* degree polynomial and knot vector (Ξ) being selected, we need to prepare the corresponding number of control points (CP) required. The increasing of control points being used to construct the NURBS curve will increase the degree-of-freedom of the beam as illustrated in Fig. 4.8. The figure shows the illustration of degree-of-freedom at each control points at the beam only when the general displacements have the same degree of polynomial.

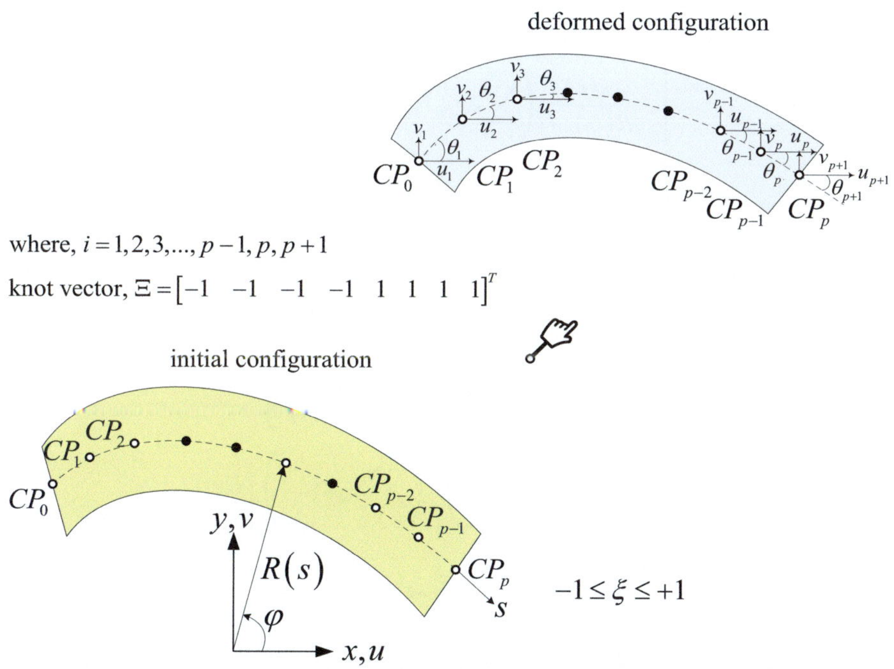

Fig. 4.8 Conceptual kinematic of general curved Timoshenko beam element in isogeometric approach

The equilibrium equation is obtained by minimizing the potential energy, given as

$$\begin{aligned}\Pi &= \frac{1}{2}\mathbf{d}^T\mathbf{K}\mathbf{d} - \frac{1}{2}\ddot{\mathbf{d}}^T\mathbf{M}\ddot{\mathbf{d}} - \mathbf{d}^T\mathbf{f} \\ \frac{\partial \Pi}{\partial \mathbf{d}^T} &= \mathbf{K}\mathbf{d} - \mathbf{M}\ddot{\mathbf{d}} - \mathbf{f} = \mathbf{0}\end{aligned} \tag{4.25}$$

Referring to (3.173), the stiffness matrix is given as

$$\begin{aligned}
\mathbf{K} &= \begin{bmatrix} \mathbf{k}_{11} & \mathbf{k}_{12} & \mathbf{k}_{13} \\ \mathbf{k}_{21} & \mathbf{k}_{22} & \mathbf{k}_{23} \\ \mathbf{k}_{31} & \mathbf{k}_{32} & \mathbf{k}_{33} \end{bmatrix} \\
\mathbf{k}_{11} &= \sum_{j=1}^{ngu} \begin{bmatrix} \frac{EA}{J_j^2}\left(\sum_{i=0}^{nu} S_{i,pu}^1(\xi_j)\right)^T \left(\sum_{i=0}^{nu} S_{i,pu}^1(\xi_j)\right) \\ +\frac{\kappa GA}{R(\xi_j)^2}\left(\sum_{i=0}^{nv} S_{i,pv}(\xi_j)\right)^T \left(\sum_{i=0}^{nv} S_{i,pv}(\xi_j)\right) \end{bmatrix} Det(J_j)\ w_j \\
\mathbf{k}_{12} = \mathbf{k}_{21} &= \sum_{j=1}^{ngu} \begin{bmatrix} \frac{EA}{R(\xi_j)\ J_j}\left(\sum_{i=0}^{nu} S_{i,pu}^1(\xi_j)\right)^T \left(\sum_{i=0}^{nv} S_{i,pv}(\xi_j)\right) \\ -\frac{\kappa GA}{R(\xi_j)\ J_j}\left(\sum_{i=0}^{nu} S_{i,pu}(\xi_j)\right)^T \left(\sum_{i=0}^{nv} S_{i,pv}^1(\xi_j)\right) \end{bmatrix} Det(J_j)\ w_j \\
\mathbf{k}_{13} = \mathbf{k}_{31} &= \sum_{j=1}^{ngu} \left[\frac{\kappa GA}{R(\xi_j)}\left(\sum_{i=0}^{nu} S_{i,pu}(\xi_j)\right)^T \left(\sum_{i=0}^{n\theta} S_{i,p\theta}(\xi_j)\right)\right] Det(J_j)\ w_j \\
\mathbf{k}_{22} &= \sum_{j=1}^{ngv} \begin{bmatrix} \frac{\kappa GA}{J_j^2}\left(\sum_{i=0}^{nv} S_{i,pv}^1(\xi_j)\right)^T \left(\sum_{i=0}^{nv} S_{i,pv}^1(\xi_j)\right) \\ +\frac{EA}{R(\xi_j)^2}\left(\sum_{i=0}^{nv} S_{i,pv}(\xi_j)\right)^T \left(\sum_{i=0}^{nv} S_{i,pv}(\xi_j)\right) \end{bmatrix} Det(J_j)\ w_j \\
\mathbf{k}_{23} = \mathbf{k}_{32} &= \sum_{j=1}^{ngv} \left[\frac{\kappa GA}{J_j}\left(\sum_{i=0}^{nv} S_{i,pv}^1(\xi_j)\right)^T \left(\sum_{i=0}^{n\theta} S_{i,p\theta}(\xi_j)\right)\right] Det(J_j)\ w_j \\
\mathbf{k}_{33} &= \sum_{j=1}^{ng\theta} \begin{bmatrix} \frac{EI}{J_j^2}\left(\sum_{i=0}^{n\theta} S_{i,p\theta}^1(\xi_j)\right)^T \left(\sum_{i=0}^{n\theta} S_{i,p\theta}^1(\xi_j)\right) \\ +\kappa GA\left(\sum_{i=0}^{n\theta} S_{i,p\theta}(\xi_j)\right)^T \left(\sum_{i=0}^{n\theta} S_{i,p\theta}(\xi_j)\right) \end{bmatrix} Det(J_j)\ w_j
\end{aligned} \tag{4.26}$$

The displacements and acceleration vectors are given as

$$\begin{aligned}
\mathbf{u} &= \begin{bmatrix} u_i & . & . & . & u_{i=pu+1} \end{bmatrix}^T \quad ; \quad \ddot{\mathbf{u}} = \begin{bmatrix} \ddot{u}_i & . & . & . & \ddot{u}_{i=pu+1} \end{bmatrix}^T \\
\mathbf{v} &= \begin{bmatrix} v_i & . & . & . & v_{i=pv+1} \end{bmatrix}^T \quad ; \quad \ddot{\mathbf{v}} = \begin{bmatrix} \ddot{v}_i & . & . & . & \ddot{v}_{i=pv+1} \end{bmatrix}^T \\
\boldsymbol{\theta} &= \begin{bmatrix} \theta_i & . & . & . & \theta_{i=p\theta+1} \end{bmatrix}^T \quad ; \quad \ddot{\boldsymbol{\theta}} = \begin{bmatrix} \ddot{\theta}_i & . & . & . & \ddot{\theta}_{i=p\theta+1} \end{bmatrix}^T
\end{aligned} \tag{4.27}$$

Referring to (3.174), the mass matrix is given by

$$\begin{aligned}
\mathbf{M} &= \begin{bmatrix} \mathbf{m}_{11} & 0 & 0 \\ 0 & \mathbf{m}_{22} & 0 \\ 0 & 0 & \mathbf{m}_{33} \end{bmatrix} \\
\mathbf{m}_{11} &= \sum_{j=1}^{ngu} \left[\rho A \left(\sum_{i=0}^{nu} S_{i,pu}(\xi_j) \right)^T \left(\sum_{i=0}^{nu} S_{i,pu}(\xi_j) \right) \right] Det(J_j) \ w_j \\
\mathbf{m}_{22} &= \sum_{j=1}^{ngv} \left[\rho A \left(\sum_{i=0}^{nv} S_{i,pv}(\xi_j) \right)^T \left(\sum_{i=0}^{nv} S_{i,pv}(\xi_j) \right) \right] Det(J_j) \ w_j \\
\mathbf{m}_{33} &= \sum_{j=1}^{ng\theta} \left[\rho I \left(\sum_{i=0}^{n\theta} S_{i,p\theta}(\xi_j) \right)^T \left(\sum_{i=0}^{n\theta} S_{i,p\theta}(\xi_j) \right) \right] Det(J_j) \ w_j
\end{aligned} \tag{4.28}$$

Referring to (3.175), the loading vector is given by

$$\mathbf{f} = \begin{Bmatrix} \mathbf{P} \\ \mathbf{Q} \\ \mathbf{M} \end{Bmatrix} + \begin{Bmatrix} \sum_{j=1}^{ngu} \left[\left(\sum_{i=0}^{nu} S_{i,pu}(\xi_j) \right) p \right] Det(J_j) \ w_j \\ \sum_{j=1}^{ngv} \left[\left(\sum_{i=0}^{nv} S_{i,pv}(\xi_j) \right) q \right] Det(J_j) \ w_j \\ \sum_{j=1}^{ng\theta} \left[\left(\sum_{i=0}^{n\theta} S_{i,p\theta}(\xi_j) \right) m \right] Det(J_j) \ w_j \end{Bmatrix} \tag{4.29}$$

4.3.3 Circular Curved Timoshenko Beam

In this section, a circular curved Timoshenko beam element will be discussed. A circular curved beam is a special case of the general curved where the radius of curvature of the curve is constant along the beam axis.

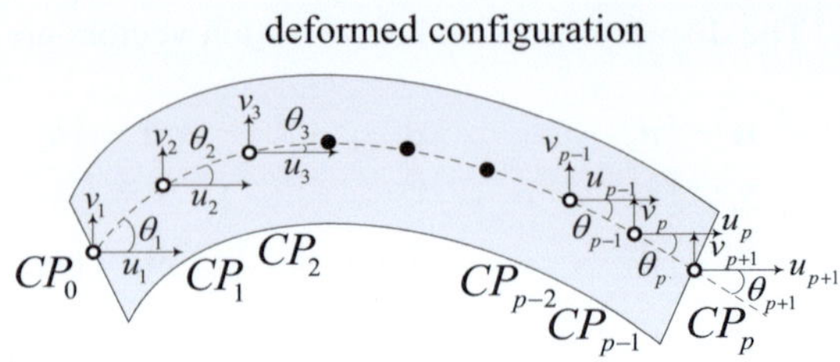

where, $i = 1, 2, 3, \ldots, p-1, p, p+1$

knot vector, $\Xi = \begin{bmatrix} -1 & -1 & -1 & -1 & 1 & 1 & 1 & 1 \end{bmatrix}^T$

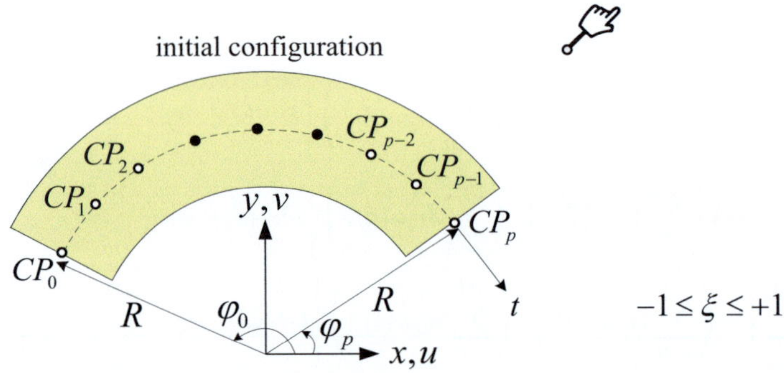

Fig. 4.9 Conceptual kinematic of circular curved Timoshenko beam element in isogeometric approach

Depending on the order of the *p-th* degree polynomial and knot vector (Ξ) being selected, we need to prepare the corresponding number of control points (CP) required. The increasing of control points being used to construct the NURBS curve will increase the degree-of-freedom of the beam as illustrated in Fig. 4.9.

The displacements and rotation of any arbitrary point along the circular curved Timoshenko beam can be calculated from the general nodal displacement of the beam element by using the rational B-spline basis functions which are given by

$$u = \sum_{i=0}^{nu} S_{i,pu}(\xi)\, u_i\,; \qquad v = \sum_{i=0}^{nv} S_{i,pv}(\xi)\, v_i\,; \qquad \theta = \sum_{i=0}^{n\theta} S_{i,p\theta}(\xi)\, \theta_i \tag{4.30}$$

By setting a constant radius of curvature, the stiffness matrix of (3.177) can be given as

$$
\mathbf{K} = \begin{bmatrix} \mathbf{k}_{11} & \mathbf{k}_{12} & \mathbf{k}_{13} \\ \mathbf{k}_{21} & \mathbf{k}_{22} & \mathbf{k}_{23} \\ \mathbf{k}_{31} & \mathbf{k}_{32} & \mathbf{k}_{33} \end{bmatrix}
$$

$$
\mathbf{k}_{11} = \sum_{j=1}^{ngu} \begin{bmatrix} \dfrac{EA}{J_j^2}\left(\sum_{i=0}^{nu} S_{i,pu}^1(\xi_j)\right)^T \left(\sum_{i=0}^{nu} S_{i,pu}^1(\xi_j)\right) \\ +\dfrac{\kappa GA}{R^2}\left(\sum_{i=0}^{nv} S_{i,pv}(\xi_j)\right)^T \left(\sum_{i=0}^{nv} S_{i,pv}(\xi_j)\right) \end{bmatrix} Det(J_j)\ w_j
$$

$$
\mathbf{k}_{12} = \mathbf{k}_{21} = \sum_{j=1}^{ngu} \begin{bmatrix} \dfrac{EA}{R\ J_j}\left(\sum_{i=0}^{nu} S_{i,pu}^1(\xi_j)\right)^T \left(\sum_{i=0}^{nv} S_{i,pv}(\xi_j)\right) \\ -\dfrac{\kappa GA}{R\ J_j}\left(\sum_{i=0}^{nu} S_{i,pu}(\xi_j)\right)^T \left(\sum_{i=0}^{nv} S_{i,pv}^1(\xi_j)\right) \end{bmatrix} Det(J_j)\ w_j
$$

$$
\mathbf{k}_{13} = \mathbf{k}_{31} = \sum_{j=1}^{ngu} \left[\frac{\kappa GA}{R}\left(\sum_{i=0}^{nu} S_{i,pu}(\xi_j)\right)^T \left(\sum_{i=0}^{n\theta} S_{i,p\theta}(\xi_j)\right)\right] Det(J_j)\ w_j \tag{4.31}
$$

$$
\mathbf{k}_{22} = \sum_{j=1}^{ngv} \begin{bmatrix} \dfrac{\kappa GA}{J_j^2}\left(\sum_{i=0}^{nv} S_{i,pv}^1(\xi_j)\right)^T \left(\sum_{i=0}^{nv} S_{i,pv}^1(\xi_j)\right) \\ +\dfrac{EA}{R^2}\left(\sum_{i=0}^{nv} S_{i,pv}(\xi_j)\right)^T \left(\sum_{i=0}^{nv} S_{i,pv}(\xi_j)\right) \end{bmatrix} Det(J_j)\ w_j
$$

$$
\mathbf{k}_{23} = \mathbf{k}_{32} = \sum_{j=1}^{ngv} \left[\frac{\kappa GA}{J_j}\left(\sum_{i=0}^{nv} S_{i,pv}^1(\xi_j)\right)^T \left(\sum_{i=0}^{n\theta} S_{i,p\theta}(\xi_j)\right)\right] Det(J_j)\ w_j
$$

$$
\mathbf{k}_{33} = \sum_{j=1}^{ng\theta} \begin{bmatrix} \dfrac{EI}{J_j^2}\left(\sum_{i=0}^{n\theta} S_{i,p\theta}^1(\xi_j)\right)^T \left(\sum_{i=0}^{n\theta} S_{i,p\theta}^1(\xi_j)\right) \\ +\kappa GA\left(\sum_{i=0}^{n\theta} S_{i,p\theta}(\xi_j)\right)^T \left(\sum_{i=0}^{n\theta} S_{i,p\theta}(\xi_j)\right) \end{bmatrix} Det(J_j)\ w_j
$$

The displacements and acceleration vectors are given as

$$
\begin{aligned}
\mathbf{u} &= \begin{bmatrix} u_i & . & . & . & u_{i=pu+1} \end{bmatrix}^T \quad ; \quad \ddot{\mathbf{u}} = \begin{bmatrix} \ddot{u}_i & . & . & . & \ddot{u}_{i=pu+1} \end{bmatrix}^T \\
\mathbf{v} &= \begin{bmatrix} v_i & . & . & . & v_{i=pv+1} \end{bmatrix}^T \quad ; \quad \ddot{\mathbf{v}} = \begin{bmatrix} \ddot{v}_i & . & . & . & \ddot{v}_{i=pv+1} \end{bmatrix}^T \\
\boldsymbol{\theta} &= \begin{bmatrix} \theta_i & . & . & . & \theta_{i=p\theta+1} \end{bmatrix}^T \quad ; \quad \ddot{\boldsymbol{\theta}} = \begin{bmatrix} \ddot{\theta}_i & . & . & . & \ddot{\theta}_{i=p\theta+1} \end{bmatrix}^T
\end{aligned} \tag{4.32}
$$

Referring to (3.178), the mass matrix is given by

$$\mathbf{M} = \begin{bmatrix} \mathbf{m}_{11} & 0 & 0 \\ 0 & \mathbf{m}_{22} & 0 \\ 0 & 0 & \mathbf{m}_{33} \end{bmatrix}$$

$$\begin{aligned}
\mathbf{m}_{11} &= \sum_{j=1}^{ngu} \left[\rho A \left(\sum_{i=0}^{nu} S_{i,pu}(\xi_j) \right)^T \left(\sum_{i=0}^{nu} S_{i,pu}(\xi_j) \right) \right] Det(J_j)\ w_j \\
\mathbf{m}_{22} &= \sum_{j=1}^{ngv} \left[\rho A \left(\sum_{i=0}^{nv} S_{i,pv}(\xi_j) \right)^T \left(\sum_{i=0}^{nv} S_{i,pv}(\xi_j) \right) \right] Det(J_j)\ w_j \\
\mathbf{m}_{33} &= \sum_{j=1}^{ng\theta} \left[\rho I \left(\sum_{i=0}^{n\theta} S_{i,p\theta}(\xi_j) \right)^T \left(\sum_{i=0}^{n\theta} S_{i,p\theta}(\xi_j) \right) \right] Det(J_j)\ w_j
\end{aligned} \tag{4.33}$$

Referring to (3.179), the loading vector is given by

$$\mathbf{f} = \begin{Bmatrix} \mathbf{P} \\ \mathbf{Q} \\ \mathbf{M} \end{Bmatrix} + \begin{Bmatrix} \sum_{j=1}^{ngu} \left[\left(\sum_{i=0}^{nu} S_{i,pu}(\xi_j) \right) p \right] Det(J_j)\ w_j \\ \sum_{j=1}^{ngv} \left[\left(\sum_{i=0}^{nv} S_{i,pv}(\xi_j) \right) q \right] Det(J_j)\ w_j \\ \sum_{j=1}^{ng\theta} \left[\left(\sum_{i=0}^{n\theta} S_{i,p\theta}(\xi_j) \right) m \right] Det(J_j)\ w_j \end{Bmatrix} \tag{4.34}$$

4.3.4 KMmatrixCircTiNURBS Program List

MATLAB code KMmatrixCircTiNURBS.m and functions: Legendre.m (Sect. 2.3.2) and DNurbsLeibnitz.m (Sect. 1.3.14) integrate numerically the stiffness and mass matrices of a circular curved Timoshenko beam example by using NURBS functions. The computed matrices of beam example in Fig. 4.10 are shown in Fig. 4.11.

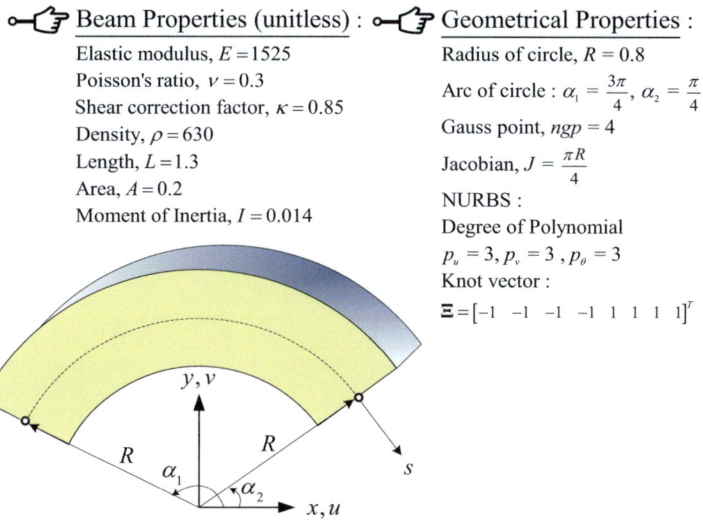

Fig. 4.10 Circular curved Timoshenko beam example

```
% KMmatrixCircTiNURBS.m
% Construct matrices K and M
% Based on the Timoshenko Beam Theory by using NURBS shape
functions
clear all;clc;
E   = 1525; A = 0.2; I = 0.014; kp = 0.85; nu = 0.3;
rho = 630; G = E/2/(1+nu);  R = 0.8;  alp1 = 3*pi/4; alp2 = pi/4;
% NURBS for u displacement data
npa   = 3;
knota  = [-ones(1,npa+1) ones(1,npa+1)]; knota = [knota ones(1,
npa)];
nba   = size(knota,2)-npa-1;
wta   = ones(nba-npa);
% NURBS for v displacement data
npb   = 3;
knotb  = [-ones(1,npb+1) ones(1,npb+1)]; knotb = [knotb ones(1,
npb)];
nbb   = size(knotb,2)-npb-1;
wtb   = ones(nbb-npb);
% NURBS for th displacement data
npc   = 3;
knotc  = [-ones(1,npc+1) ones(1,npc+1)]; knotc = [knotc ones(1,
npc)];
nbc   = size(knotc,2)-npc-1;
wtc   = ones(nbc-npc);
```

The Stiffness Matrix

464.8493	-204.4556	-140.0330	-71.4149	-128.3053	-151.7668	-75.8834	-25.2945	22.3752	11.1876	4.4750	1.1188
-204.4556	308.0350	85.3994	-140.0330	151.7668	-0.0000	-75.8834	-75.8834	11.1876	13.4251	10.0688	4.4750
-140.0330	85.3994	308.0350	-204.4556	75.8834	75.8834	0.0000	-151.7668	4.4750	10.0688	13.4251	11.1876
-71.4149	-140.0330	-204.4556	464.8493	25.2945	75.8834	151.7668	128.3053	1.1188	4.4750	11.1876	22.3752
-128.3053	151.7668	75.8834	25.2945	228.3786	-28.6370	-30.4983	-19.5268	49.8558	29.9135	14.9567	4.9856
-151.7668	-0.0000	75.8834	75.8834	-28.6370	146.5489	62.3029	-30.4983	-29.9135	0.0000	14.9567	14.9567
-75.8834	-75.8834	0.0000	151.7668	-30.4983	62.3029	146.5489	-28.6370	-14.9567	-14.9567	0.0000	29.9135
-25.2945	-75.8834	-151.7668	128.3053	-19.5268	-30.4983	-28.6370	228.3786	-4.9856	-14.9567	-29.9135	-49.8558
22.3752	11.1876	4.4750	1.1188	49.8558	-29.9135	-14.9567	-4.9856	48.4818	-6.3407	-6.6138	-4.2019
11.1876	13.4251	10.0688	4.4750	29.9135	0.0000	-14.9567	-14.9567	-6.3407	31.1279	13.1520	-6.6138
4.4750	10.0688	13.4251	11.1876	14.9567	14.9567	0.0000	-29.9135	-6.6138	13.1520	31.1279	-6.3407
1.1188	4.4750	11.1876	22.3752	4.9856	14.9567	29.9135	-49.8558	-4.2019	-6.6138	-6.3407	48.4818

The Mass Matrix

22.6195	11.3097	4.5239	1.1310	0	0	0	0	0	0	0	0
11.3097	13.5717	10.1788	4.5239	0	0	0	0	0	0	0	0
4.5239	10.1788	13.5717	11.3097	0	0	0	0	0	0	0	0
1.1310	4.5239	11.3097	22.6195	0	0	0	0	0	0	0	0
0	0	0	0	22.6195	11.3097	4.5239	1.1310	0	0	0	0
0	0	0	0	11.3097	13.5717	10.1788	4.5239	0	0	0	0
0	0	0	0	4.5239	10.1788	13.5717	11.3097	0	0	0	0
0	0	0	0	1.1310	4.5239	11.3097	22.6195	0	0	0	0
0	0	0	0	0	0	0	0	1.5834	0.7917	0.3167	0.0792
0	0	0	0	0	0	0	0	0.7917	0.9500	0.7125	0.3167
0	0	0	0	0	0	0	0	0.3167	0.7125	0.9500	0.7917
0	0	0	0	0	0	0	0	0.0792	0.3167	0.7917	1.5834

Fig. 4.11 The stiffness and mass matrices of a circular curved Timoshenko beam element example

```
% Jacobi and Gauss
Jac   = abs(alp2-alp1)/2*R; % Constant for a circular arc
ngp = 4;   Gpw   = LegendreN(ngp);
dim   = npa+1+npb+1+npc+1;
```

```
Kmtx = zeros(dim,dim);      Mmtx = zeros(dim,dim);
Su0  = zeros(nba-npa,ngp); Sv0  = zeros(nbb-npb,ngp);
St0  = zeros(nbc-npc,ngp); Su1  = zeros(nba-npa,ngp);
Sv1  = zeros(nbb-npb,ngp); St1  = zeros(nbc-npc,ngp);
for i=1:ngp
  t = Gpw(i,1); % coordinate
  DNurbsbasis = DNurbsLeibnitz(npa,knota,wta,t);
  Su0(1:nba-npa,i)=DNurbsbasis(1:nba-npa,npa+5); % non-derivative
  Su1(1:nba-npa,i)=DNurbsbasis(1:nba-npa,npa+7); % 1st-derivative
end
for i=1:ngp
  t = Gpw(i,1); % coordinate
  DNurbsbasis = DNurbsLeibnitz(npb,knotb,wtb,t);
  Sv0(1:nbb-npb,i)=DNurbsbasis(1:nbb-npb,npb+5); % non-derivative
  Sv1(1:nbb-npb,i)=DNurbsbasis(1:nbb-npb,npb+7); % 1st-derivative
end
for i=1:ngp
  t = Gpw(i,1); % coordinate
  DNurbsbasis = DNurbsLeibnitz(npc,knotc,wtc,t);
  St0(1:nbc-npc,i)=DNurbsbasis(1:nbc-npc,npc+5); % non-derivative
  St1(1:nbc-npc,i)=DNurbsbasis(1:nbc-npc,npc+7); % 1st-derivative
end
% K11
nna = nba-npa;
for i=1:ngp
  for j=1:nna
    for k=1:nna
      Kmtx(j,k) = Kmtx(j,k)+E*A*Su1(j,i)/Jac*Su1(k,i)/Jac*Jac*Gpw
      (i,2)...
          +kp*G*A*Su0(j,i)/R*Su0(k,i)/R*Jac*Gpw(i,2);
    end
  end
end
% K12
nnb = nbb-npb;
for i=1:ngp
  for j=1:nna
    for k=1:nnb
      Kmtx(j,k+nna) = Kmtx(j,k+nna)...
          +E*A*Su1(j,i)/Jac*Sv0(k,i)/R*Jac*Gpw(i,2)...
          -kp*G*A*Su0(j,i)/R*Sv1(k,i)/Jac*Jac*Gpw(i,2);
      Kmtx(k+nna,j) = Kmtx(j,k+nna); % Symmetric K21
    end
  end
end
```

```
% K13
nnc = nbc-npc;
for i=1:ngp
  for j=1:nna
    for k=1:nnc
      Kmtx(j,k+nna+nnb) = Kmtx(j,k+nna+nnb)...
          +kp*G*A*Su0(j,i)/R*St0(k,i)*Jac*Gpw(i,2);
      Kmtx(k+nna+nnb,j) = Kmtx(j,k+nna+nnb); % Symmetric K31
    end
  end
end
% K22
for i=1:ngp
  for j=1:nnb
    for k=1:nnb
      Kmtx(j+nna,k+nna) = Kmtx(j+nna,k+nna)...
          +E*A*Sv0(j,i)/R*Sv0(k,i)/R*Jac*Gpw(i,2)...
          +kp*G*A*Sv1(j,i)/Jac*Sv1(k,i)/Jac*Jac*Gpw(i,2);
    end
  end
end
% K23
for i=1:ngp
  for j=1:nnb
    for k=1:nnc
      Kmtx(j+nna,k+nna+nnb) = Kmtx(j+nna,k+nna+nnb)...
          -kp*G*A*Sv1(j,i)/Jac*St0(k,i)*Jac*Gpw(i,2);
       Kmtx(k+nna+nnb,j+nna) = Kmtx(j+nna,k+nna+nnb); % Symmetric
       K32
    end
  end
end
% K33
for i=1:ngp
  for j=1:nnc
    for k=1:nnc
      Kmtx(j+nna+nnb,k+nna+nnb) = Kmtx(j+nna+nnb,k+nna+nnb)...
          +kp*G*A*St0(j,i)*St0(k,i)*Jac*Gpw(i,2)...
          +E*I*St1(j,i)/Jac*St1(k,i)/Jac*Jac*Gpw(i,2);
    end
  end
end
disp('The Stiffness Matrix');
disp(Kmtx);
% Mass matrix K
```

```
% M11
for i=1:ngp
  for j=1:nna
    for k=1:nna
      Mmtx(j,k) = Mmtx(j,k)+rho*A*Su0(j,i)*Su0(k,i)*Jac*Gpw(i,2);
    end
  end
end
% M22
for i=1:ngp
  for j=1:nnb
    for k=1:nnb
      Mmtx(j+nna,k+nna) = Mmtx(j+nna,k+nna)...
          +rho*A*Sv0(j,i)*Sv0(k,i)*Jac*Gpw(i,2);
    end
  end
end
% M33
for i=1:ngp
  for j=1:nnc
    for k=1:nnc
      Mmtx(j+nna+nnb,k+nna+nnb) = Mmtx(j+nna+nnb,k+nna+nnb)...
          +rho*I*St0(j,i)*St0(k,i)*Jac*Gpw(i,2);
    end
  end
end
disp('The Mass Matrix');
disp(Mmtx);
```

We can see from Fig. 4.11 that the computed stiffness and mass matrices have twelve degree-of-freedoms beam element, since the horizontal displacement, u, v, and rotational θ were assumed to be third order of NURBS. The stiffness and mass matrices of the element will be condensed and rearranged to suit the standard two-node beam element with six degree-of-freedoms.

References

Basilevs Y, Calo VM, Cottrell JA, Evans J, Hughes TJR, Lipton S, Scott MA, Sederberg TW (2010) Isogeometric analysis using T-splines. Comput Meth Appl Mech Engrg 199:229–263

Cottrell JA, Reali A, Bazilevs Y, Hughes TJR (2006) Isogeometric analysis of structural vibrations. Comput Meth Appl Mech Engrg 195:5257–5296

Cottrell JA, Hughes TJR, Reali A (2007) Studies of refinement and continuity in isogeometric analysis. Comput Meth Appl Mech Engrg 196:4160–4183

Gontier C, Vollmer C (1995) A large displacement analysis of a beam using a CAD geometric definition. Comput Struct 57(6):981–989
Schramm U, Pilkey WD (1993) The coupling of geometric descriptions and finite elements using NURBs – a study in shape optimization. Finite Elem Anal Des 15:11–34

Further Reading

Cottrell JA, TJR H, Bazilevs Y (2009) Isogeometric analysis: toward integration of CAD and FEA. Wiley, Chichester

Chapter 5
Condensation Method

5.1 Introduction

The traditional static condensation and dynamic condensation is known to be effective in reducing the degree of freedom of a beam which has more than two nodes. The amounts of degree of freedom of the mid-nodes are "hidden" so that only the degrees of freedom at both ends of the beam still retained (Bathe 1982; Cook 1981; Hart and Wong 2000).

Although the importance of condensation is not significant in beam problems because the amounts of degree of freedom are not vast, condensing numerous degrees of freedom to the standard two-node six degree of freedom beam element has a lot of benefits for real practices. First, the introduction of NURBS into the design practice will hinder some design practitioners that are not familiar with the new concept. Second, the condensed NURBS element codes can be integrated seamlessly into the existing beam finite element codes.

In the conventional static and dynamic condensation methods, the beam element matrix is usually partitioned into two sub-matrices which consist of retained and condensed parts. Therefore, there are only two simultaneous equations to solve, that is, a linear relationship of two unknowns. From there, the condensed part will be eliminated from the equations; hence the modified retained part is used to solve the beam problems. This concept is very suitable for the NURBS functions where the degree of freedom of the mid-nodes increased when using high-order degree of the polynomial.

B.S. Gan, *An Isogeometric Approach to Beam Structures*,
DOI 10.1007/978-3-319-56493-7_5

5.2 Proposed Condensation Method

In present proposed condensation method, the element matrices are breakdown to three sub-matrix simultaneous equations, where the mid-node degrees of freedom are subjected to condensation. We may call this condensation as *Sandwich* Condensation.

By breaking down the matrix into three simultaneous sub-matrix equations, we can obtain a second-order relationship between the general displacements, because there are three unknowns. The conventional dynamic condensation is then modified by taking twice derivatives to the free vibration equilibrium equations against the time *t*. Then, we can obtain a relationship between the stiffness matrix **k** and the acceleration vector $\ddot{\mathbf{d}}$.

5.3 Static Condensation

We start from a static equilibrium of three simultaneous equations which is represented in the matrix form:

$$\begin{bmatrix} \mathbf{k}_{11} & \mathbf{k}_{12} & \mathbf{k}_{13} \\ \mathbf{k}_{21} & \mathbf{k}_{22} & \mathbf{k}_{23} \\ \mathbf{k}_{31} & \mathbf{k}_{32} & \mathbf{k}_{33} \end{bmatrix} \begin{Bmatrix} \mathbf{d}_1 \\ \mathbf{d}_2 \\ \mathbf{d}_3 \end{Bmatrix} = \begin{Bmatrix} \mathbf{f}_1 \\ \mathbf{f}_2 \\ \mathbf{f}_3 \end{Bmatrix} \tag{5.1}$$

We can have three simultaneous equations as follows:

$$\begin{aligned} &(\mathrm{a})\ \mathbf{k}_{11}\mathbf{d}_1 + \mathbf{k}_{12}\mathbf{d}_2 + \mathbf{k}_{13}\mathbf{d}_3 = \mathbf{f}_1 \\ &(\mathrm{b})\ \mathbf{k}_{21}\mathbf{d}_1 + \mathbf{k}_{22}\mathbf{d}_2 + \mathbf{k}_{23}\mathbf{d}_3 = \mathbf{f}_2 \\ &(\mathrm{c})\ \mathbf{k}_{31}\mathbf{d}_1 + \mathbf{k}_{32}\mathbf{d}_2 + \mathbf{k}_{33}\mathbf{d}_3 = \mathbf{f}_3 \end{aligned}$$

Extracting $\mathbf{d}_2$ from (b)

$$\begin{aligned} \mathbf{k}_{22}\mathbf{d}_2 &= \mathbf{f}_2 - \mathbf{k}_{21}\mathbf{d}_1 - \mathbf{k}_{23}\mathbf{d}_3 \\ \mathbf{d}_2 &= \mathbf{k}_{22}^{-1}(\mathbf{f}_2 - \mathbf{k}_{21}\mathbf{d}_1 - \mathbf{k}_{23}\mathbf{d}_3) \end{aligned}$$

Substituting $\mathbf{d}_2$ into (a)

$$\mathbf{k}_{11}\mathbf{d}_1 + \mathbf{k}_{12}\mathbf{k}_{22}^{-1}(\mathbf{f}_2 - \mathbf{k}_{21}\mathbf{d}_1 - \mathbf{k}_{23}\mathbf{d}_3) + \mathbf{k}_{13}\mathbf{d}_3 = \mathbf{f}_1$$

$$\left(\mathbf{k}_{11} - \mathbf{k}_{12}\mathbf{k}_{22}^{-1}\mathbf{k}_{21}\right)\mathbf{d}_1 + \left(\mathbf{k}_{13} - \mathbf{k}_{12}\mathbf{k}_{22}^{-1}\mathbf{k}_{23}\right)\mathbf{d}_3 = \mathbf{f}_1 - \mathbf{k}_{12}\mathbf{k}_{22}^{-1}\mathbf{f}_2$$

$$\bar{\mathbf{k}}_{11}\mathbf{d}_1 + \bar{\mathbf{k}}_{13}\mathbf{d}_3 = \bar{\mathbf{f}}_1 \tag{5.2}$$

where

$$\begin{aligned}
\bar{\mathbf{k}}_{11} &= \mathbf{k}_{11} - \mathbf{k}_{12}\mathbf{k}_{22}^{-1}\mathbf{k}_{21} \\
\bar{\mathbf{k}}_{13} &= \mathbf{k}_{13} - \mathbf{k}_{12}\mathbf{k}_{22}^{-1}\mathbf{k}_{23} \\
\bar{\mathbf{f}}_{1} &= \mathbf{f}_{1} - \mathbf{k}_{12}\mathbf{k}_{22}^{-1}\mathbf{f}_{2}
\end{aligned}$$

Substituting $\mathbf{d}_2$ into (c)

$$\begin{aligned}
&\mathbf{k}_{31}\mathbf{d}_1 + \mathbf{k}_{32}\mathbf{k}_{22}^{-1}(\mathbf{f}_2 - \mathbf{k}_{21}\mathbf{d}_1 - \mathbf{k}_{23}\mathbf{d}_3) + \mathbf{k}_{33}\mathbf{d}_3 = \mathbf{f}_3 \\
&\left(\mathbf{k}_{31} - \mathbf{k}_{32}\mathbf{k}_{22}^{-1}\mathbf{k}_{21}\right)\mathbf{d}_1 + \left(\mathbf{k}_{33} - \mathbf{k}_{32}\mathbf{k}_{22}^{-1}\mathbf{k}_{23}\right)\mathbf{d}_3 = \mathbf{f}_3 - \mathbf{k}_{32}\mathbf{k}_{22}^{-1}\mathbf{f}_2 \\
&\bar{\mathbf{k}}_{31}\mathbf{d}_1 + \bar{\mathbf{k}}_{33}\mathbf{d}_3 = \bar{\mathbf{f}}_3
\end{aligned} \tag{5.3}$$

where

$$\begin{aligned}
\bar{\mathbf{k}}_{31} &= \mathbf{k}_{31} - \mathbf{k}_{32}\mathbf{k}_{22}^{-1}\mathbf{k}_{21} \\
\bar{\mathbf{k}}_{33} &= \mathbf{k}_{33} - \mathbf{k}_{32}\mathbf{k}_{22}^{-1}\mathbf{k}_{23} \\
\bar{\mathbf{f}}_{3} &= \mathbf{f}_{3} - \mathbf{k}_{32}\mathbf{k}_{22}^{-1}\mathbf{f}_{2}
\end{aligned}$$

The result of static condensation is given by

$$\begin{bmatrix} \bar{\mathbf{k}}_{11} & \bar{\mathbf{k}}_{13} \\ \bar{\mathbf{k}}_{31} & \bar{\mathbf{k}}_{33} \end{bmatrix} \begin{Bmatrix} \mathbf{d}_1 \\ \mathbf{d}_3 \end{Bmatrix} = \begin{Bmatrix} \bar{\mathbf{f}}_1 \\ \bar{\mathbf{f}}_3 \end{Bmatrix} \tag{5.4}$$

5.4 Modified Dynamic Condensation

Next, we continue the condensation to the free vibration mode; the equilibrium equation of free vibration in the matrix form is given as

$$\begin{bmatrix} \mathbf{k}_{11} & \mathbf{k}_{12} & \mathbf{k}_{13} \\ \mathbf{k}_{21} & \mathbf{k}_{22} & \mathbf{k}_{23} \\ \mathbf{k}_{31} & \mathbf{k}_{32} & \mathbf{k}_{33} \end{bmatrix} \begin{Bmatrix} \mathbf{d}_1 \\ \mathbf{d}_2 \\ \mathbf{d}_3 \end{Bmatrix} + \begin{bmatrix} \mathbf{m}_{11} & \mathbf{m}_{12} & \mathbf{m}_{13} \\ \mathbf{m}_{21} & \mathbf{m}_{22} & \mathbf{m}_{23} \\ \mathbf{m}_{31} & \mathbf{m}_{32} & \mathbf{m}_{33} \end{bmatrix} \begin{Bmatrix} \ddot{\mathbf{d}}_1 \\ \ddot{\mathbf{d}}_2 \\ \ddot{\mathbf{d}}_3 \end{Bmatrix} = \begin{Bmatrix} \mathbf{0} \\ \mathbf{0} \\ \mathbf{0} \end{Bmatrix} \tag{5.5}$$

To relate the displacement $\mathbf{d}$ and acceleration $\ddot{\mathbf{d}}$ in the equations, let us derive the Eq. (5.5) with respect to time t; two times become

$$\begin{bmatrix} \mathbf{k}_{11} & \mathbf{k}_{12} & \mathbf{k}_{13} \\ \mathbf{k}_{21} & \mathbf{k}_{22} & \mathbf{k}_{23} \\ \mathbf{k}_{31} & \mathbf{k}_{32} & \mathbf{k}_{33} \end{bmatrix} \begin{Bmatrix} \ddot{\mathbf{d}}_1 \\ \ddot{\mathbf{d}}_2 \\ \ddot{\mathbf{d}}_3 \end{Bmatrix} + \begin{bmatrix} \mathbf{m}_{11} & \mathbf{m}_{12} & \mathbf{m}_{13} \\ \mathbf{m}_{21} & \mathbf{m}_{22} & \mathbf{m}_{23} \\ \mathbf{m}_{31} & \mathbf{m}_{32} & \mathbf{m}_{33} \end{bmatrix} \begin{Bmatrix} \widetilde{\mathbf{d}}_1 \\ \widetilde{\mathbf{d}}_2 \\ \widetilde{\mathbf{d}}_3 \end{Bmatrix} = \begin{Bmatrix} \mathbf{0} \\ \mathbf{0} \end{Bmatrix} \tag{5.6}$$

Neglecting the fourth-order displacement terms

$$\begin{bmatrix} \mathbf{k}_{11} & \mathbf{k}_{12} & \mathbf{k}_{13} \\ \mathbf{k}_{21} & \mathbf{k}_{22} & \mathbf{k}_{23} \\ \mathbf{k}_{31} & \mathbf{k}_{32} & \mathbf{k}_{33} \end{bmatrix} \begin{Bmatrix} \ddot{\mathbf{d}}_1 \\ \ddot{\mathbf{d}}_2 \\ \ddot{\mathbf{d}}_3 \end{Bmatrix} = \begin{Bmatrix} \mathbf{0} \\ \mathbf{0} \\ \mathbf{0} \end{Bmatrix} \tag{5.7}$$

We can have three simultaneous equations as follows:

$$\begin{aligned} &(\mathrm{a})\ \mathbf{k}_{11}\ddot{\mathbf{d}}_1 + \mathbf{k}_{12}\ddot{\mathbf{d}}_2 + \mathbf{k}_{13}\ddot{\mathbf{d}}_3 = \mathbf{0} \\ &(\mathrm{b})\ \mathbf{k}_{21}\ddot{\mathbf{d}}_1 + \mathbf{k}_{22}\ddot{\mathbf{d}}_2 + \mathbf{k}_{23}\ddot{\mathbf{d}}_3 = \mathbf{0} \\ &(\mathrm{c})\ \mathbf{k}_{31}\ddot{\mathbf{d}}_1 + \mathbf{k}_{32}\ddot{\mathbf{d}}_2 + \mathbf{k}_{33}\ddot{\mathbf{d}}_3 = \mathbf{0} \end{aligned}$$

Extracting $\ddot{\mathbf{d}}_2$ from (b)

$$\ddot{\mathbf{d}}_2 = \mathbf{k}_{22}^{-1}\left(-\mathbf{k}_{21}\ddot{\mathbf{d}}_1 - \mathbf{k}_{23}\ddot{\mathbf{d}}_3\right) \tag{5.8}$$

Substituting $\ddot{\mathbf{d}}_2$ into (5.5) and solving for $\mathbf{d}_2$ from the second simultaneous equation

$$\begin{aligned} &\mathbf{k}_{21}\mathbf{d}_1 + \mathbf{k}_{22}\mathbf{d}_2 + \mathbf{k}_{23}\mathbf{d}_3 + \mathbf{m}_{21}\ddot{\mathbf{d}}_1 + \mathbf{m}_{22}\mathbf{k}_{22}^{-1}\left(-\mathbf{k}_{21}\ddot{\mathbf{d}}_1 - \mathbf{k}_{23}\ddot{\mathbf{d}}_3\right) \\ &+\mathbf{m}_{23}\ddot{\mathbf{d}}_3 = \mathbf{0}\mathbf{d}_2 = -\mathbf{k}_{22}^{-1}\mathbf{k}_{21}\mathbf{d}_1 - \mathbf{k}_{22}^{-1}\mathbf{k}_{23}\mathbf{d}_3 - \mathbf{k}_{22}^{-1}\mathbf{m}_{21}\ddot{\mathbf{d}}_1 \\ &+\mathbf{k}_{22}^{-1}\mathbf{m}_{22}\mathbf{k}_{22}^{-1}\left(\mathbf{k}_{21}\ddot{\mathbf{d}}_1 + \mathbf{k}_{23}\ddot{\mathbf{d}}_3\right) - \mathbf{k}_{22}^{-1}\mathbf{m}_{23}\ddot{\mathbf{d}}_3 \\ &\quad \mathbf{d}_2 = -\mathbf{k}_{22}^{-1}\mathbf{k}_{21}\mathbf{d}_1 - \mathbf{k}_{22}^{-1}\mathbf{k}_{23}\mathbf{d}_3 - \mathbf{k}_{22}^{-1}\bar{\mathbf{m}}_{21}\ddot{\mathbf{d}}_1 - \mathbf{k}_{22}^{-1}\bar{\mathbf{m}}_{23}\ddot{\mathbf{d}}_3 \end{aligned} \tag{5.9}$$

where

$$\begin{aligned} \bar{\mathbf{m}}_{21} &- \mathbf{m}_{21} - \mathbf{m}_{22}\mathbf{k}_{22}^{-1}\mathbf{k}_{21} \\ \bar{\mathbf{m}}_{23} &= \mathbf{m}_{23} - \mathbf{m}_{22}\mathbf{k}_{22}^{-1}\mathbf{k}_{23} \end{aligned}$$

Substituting $\ddot{\mathbf{d}}_2$ from (5.8) and $\mathbf{d}_2$ from (5.9) into the first simultaneous equation of (5.5) results in

$$\begin{aligned} &\mathbf{k}_{11}\mathbf{d}_1 + \mathbf{k}_{12}\left(-\mathbf{k}_{22}^{-1}\mathbf{k}_{21}\mathbf{d}_1 - \mathbf{k}_{22}^{-1}\mathbf{k}_{23}\mathbf{d}_3 - \mathbf{k}_{22}^{-1}\bar{\mathbf{m}}_{21}\ddot{\mathbf{d}}_1 - \mathbf{k}_{22}^{-1}\bar{\mathbf{m}}_{23}\ddot{\mathbf{d}}_3\right) \\ &+\mathbf{k}_{13}\mathbf{d}_3 + \mathbf{m}_{11}\ddot{\mathbf{d}}_1 + \mathbf{m}_{12}\mathbf{k}_{22}^{-1}\left(-\mathbf{k}_{21}\ddot{\mathbf{d}}_1 - \mathbf{k}_{23}\ddot{\mathbf{d}}_3\right) + \mathbf{m}_{13}\ddot{\mathbf{d}}_3 = \mathbf{0} \end{aligned}$$

leading to an equation

$$\bar{\mathbf{k}}_{11}\mathbf{d}_1 + \bar{\mathbf{k}}_{13}\mathbf{d}_3 + \bar{\mathbf{m}}_{11}\ddot{\mathbf{d}}_1 + \bar{\mathbf{m}}_{13}\ddot{\mathbf{d}}_3 = \mathbf{0} \tag{5.10}$$

where

$$\begin{aligned} \bar{\mathbf{m}}_{11} &= \mathbf{m}_{11} - \mathbf{m}_{12}\mathbf{k}_{22}^{-1}\mathbf{k}_{21} - \mathbf{k}_{12}\mathbf{k}_{22}^{-1}\bar{\mathbf{m}}_{21} \\ \bar{\mathbf{m}}_{13} &= \mathbf{m}_{13} - \mathbf{m}_{12}\mathbf{k}_{22}^{-1}\mathbf{k}_{23} - \mathbf{k}_{12}\mathbf{k}_{22}^{-1}\bar{\mathbf{m}}_{23} \end{aligned}$$

Substituting $\ddot{\mathbf{d}}_2$ from (5.8) and $\mathbf{d}_2$ from (5.9) into the third simultaneous equation of (5.5) results in

$$\mathbf{k}_{31}\mathbf{d}_1 + \mathbf{k}_{32}\left(-\mathbf{k}_{22}^{-1}\mathbf{k}_{21}\mathbf{d}_1 - \mathbf{k}_{22}^{-1}\mathbf{k}_{23}\mathbf{d}_3 - \mathbf{k}_{22}^{-1}\bar{\mathbf{m}}_{21}\ddot{\mathbf{d}}_1 - \mathbf{k}_{22}^{-1}\bar{\mathbf{m}}_{23}\ddot{\mathbf{d}}_3\right)$$
$$+\mathbf{k}_{33}\mathbf{d}_3 + \mathbf{m}_{31}\ddot{\mathbf{d}}_1 + \mathbf{m}_{32}\mathbf{k}_{22}^{-1}\left(-\mathbf{k}_{21}\ddot{\mathbf{d}}_1 - \mathbf{k}_{23}\ddot{\mathbf{d}}_3\right) + \mathbf{m}_{33}\ddot{\mathbf{d}}_3 = \mathbf{0}$$

leading to an equation

$$\bar{\mathbf{k}}_{31}\mathbf{d}_1 + \bar{\mathbf{k}}_{33}\mathbf{d}_3 + \bar{\mathbf{m}}_{31}\ddot{\mathbf{d}}_1 + \bar{\mathbf{m}}_{33}\ddot{\mathbf{d}}_3 = \mathbf{0} \tag{5.11}$$

where

$$\bar{\mathbf{m}}_{31} = \mathbf{m}_{31} - \mathbf{m}_{32}\mathbf{k}_{22}^{-1}\mathbf{k}_{21} - \mathbf{k}_{32}\mathbf{k}_{22}^{-1}\bar{\mathbf{m}}_{21}$$
$$\bar{\mathbf{m}}_{33} = \mathbf{m}_{33} - \mathbf{m}_{32}\mathbf{k}_{22}^{-1}\mathbf{k}_{23} - \mathbf{k}_{32}\mathbf{k}_{22}^{-1}\bar{\mathbf{m}}_{23}$$

Summarizing (5.10) and (5.11), we have a condensed modal equation that can be used for solving free vibration problems with only two nodes beam element.

$$\begin{bmatrix} \bar{\mathbf{k}}_{11} & \bar{\mathbf{k}}_{13} \\ \bar{\mathbf{k}}_{31} & \bar{\mathbf{k}}_{33} \end{bmatrix} \begin{Bmatrix} \mathbf{d}_1 \\ \mathbf{d}_3 \end{Bmatrix} + \begin{bmatrix} \bar{\mathbf{m}}_{11} & \bar{\mathbf{m}}_{13} \\ \bar{\mathbf{m}}_{31} & \bar{\mathbf{m}}_{33} \end{bmatrix} \begin{Bmatrix} \ddot{\mathbf{d}}_1 \\ \ddot{\mathbf{d}}_3 \end{Bmatrix} = \begin{Bmatrix} \mathbf{0} \\ \mathbf{0} \end{Bmatrix} \tag{5.12}$$

Following the same procedure, consecutively, we can establish the condensation method for free vibration with damping and/or obtain the dynamic condensation to include the external dynamic loading.

The present condensation method is not like the conventional modal superposition method where only the dominant modes are considered and the higher order displacements are eliminated from the formulation. In the present condensation method the "*hidden*" acceleration $\left[\ddot{\mathbf{d}}_2\right]$ and displacement $\ddot{\mathbf{d}}_2$ vectors can be recovered back by using (5.8) and (5.1) equations.

5.5 Procedure of Sandwich Condensation

The element matrix of the beam constructed by using the NURBS functions was initially arranged in each general displacement order of polynomial. We can see the beam element matrices which have not been condensed as shown in Figs. 4.6 and 4.11. The procedure of the Sandwich Condensation is shown in Fig. 5.1. In the procedure, the condensation is conducted by condensing **u**, **v**, and $\boldsymbol{\theta}$ series of matrix equations successively.

Illustratively, the processes of three stages condensations are given in Figs. 5.2 and 5.3. The sub-matrices are constructed in every stage by using (5.12), repeatedly for **u**, **v**, and $\boldsymbol{\theta}$ series of matrix equations. The final condensed matrix will be a six degree of freedom beam element matrix.

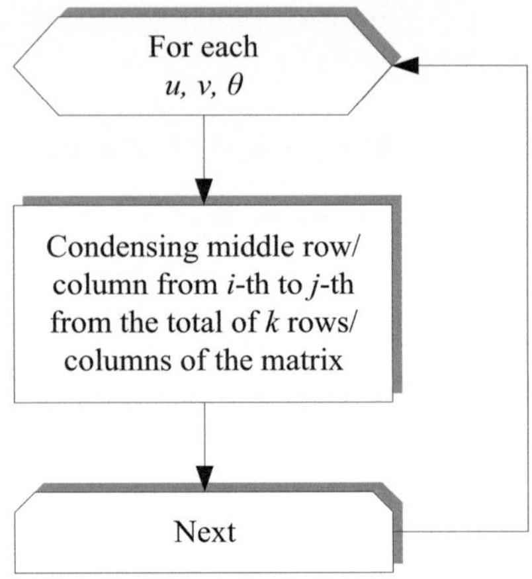

Fig. 5.1 Flowchart of the Sandwich Condensation method

5.5.1 KMmatrixTiNURBSCondensed Program List

MATLAB code KMmatrixTiNURBSCondensed.m and functions Condensation.m (Sect. 5.5.2), Legendre.m (Sect. 2.3.2), and DNurbsLeibnitz.m (Sect. 1.3.14) integrate the stiffness and mass matrices of a straight Timoshenko beam numerically by using NURBS functions. Before and after the condensation, the stiffness and mass matrices are exactly showing the same value with the results shown in Fig. 4.6 (before the condensation) and Fig. 3.14 (after the condensation).

```
% KMmatrixTiNURBSCondensed.m
% Construct matrices K and M
% Based on the Timoshenko Beam Theory by using NURBS shape functions
clear all;clc;
L = 1.3; E = 1525; A = 0.2; nu = 0.3;
I = 0.014; kp = 0.85; rho = 630;
G = E/2/(1+nu);
% NURBS for u displacement data
npa = 1;
knota = [-ones(1,npa+1) ones(1,npa+1)]; knota = [knota ones(1,npa)];
nba = size(knota,2)-npa-1;
wta = ones(nba-npa);
% NURBS for v displacement data
npb = 3;
knotb = [-ones(1,npb+1) ones(1,npb+1)]; knotb = [knotb ones(1,npb)];
nbb = size(knotb,2)-npb-1;
wtb = ones(nbb-npb);
% NURBS for th displacement data
npc = 2;
knotc = [-ones(1,npc+1) ones(1,npc+1)]; knotc = [knotc ones(1,npc)];
nbc = size(knotc,2)-npc-1;
```

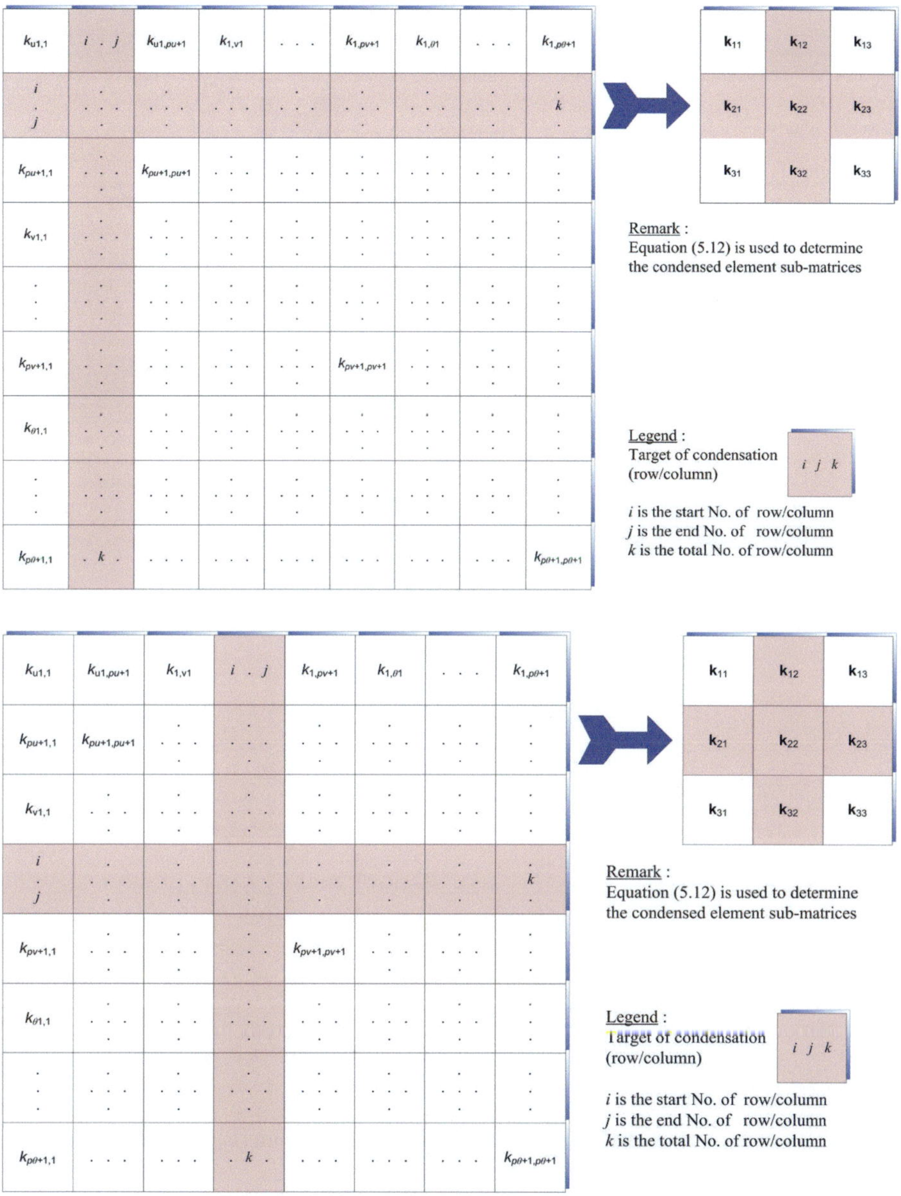

Fig. 5.2 Illustration of the Sandwich Condensation procedures: first and second stages

```
wtc = ones(nbc-npc);
% Jacobi and Gauss
Jac = L/2;
ngp = 4; Gpw = Legendre(ngp);
dim = npa+1+npb+1+npc+1;
```

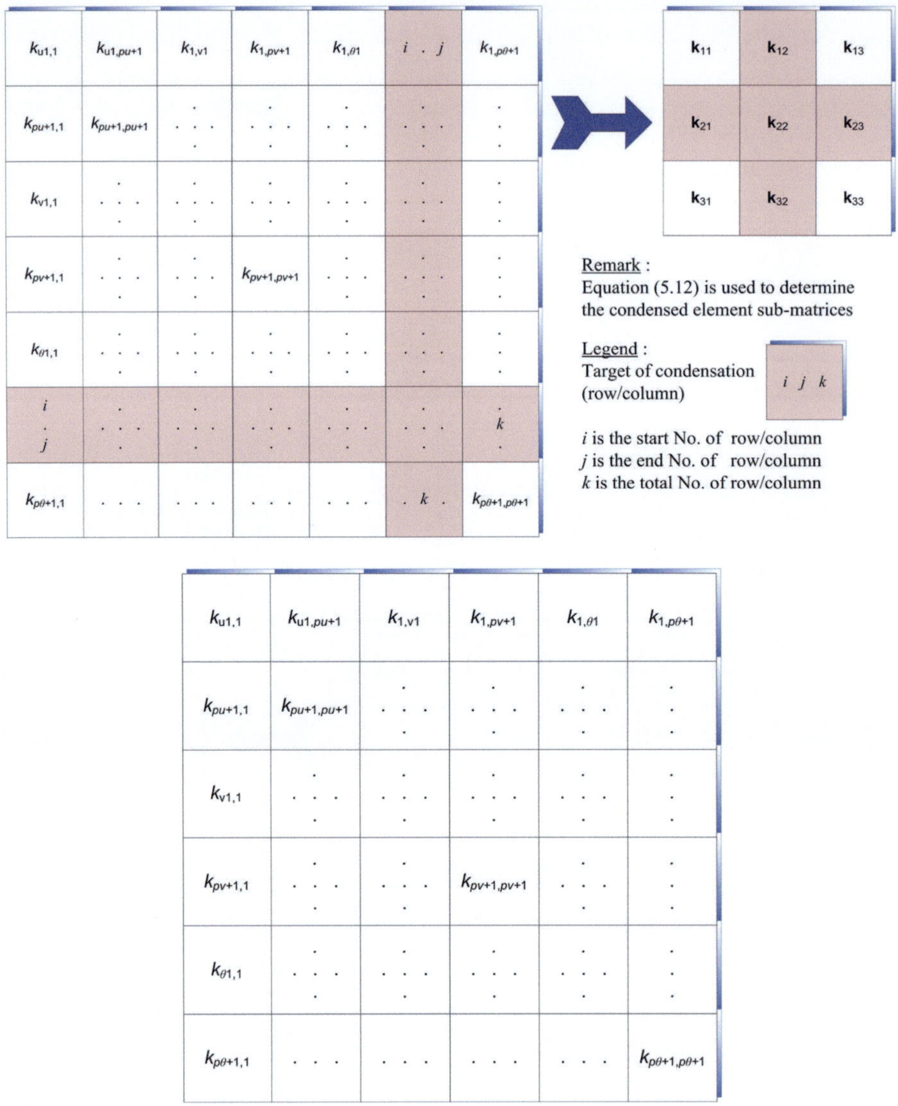

Fig. 5.3 Illustration of the Sandwich Condensation procedures: third stage and result

```
Kmtx = zeros(dim,dim); Mmtx = zeros(dim,dim);
Su0 = zeros(nba-npa,ngp); Sv0 = zeros(nbb-npb,ngp);
St0 = zeros(nbc-npc,ngp); Su1 = zeros(nba-npa,ngp);
Sv1 = zeros(nbb-npb,ngp); St1 = zeros(nbc-npc,ngp);
for i=1:ngp
   t = Gpw(i,1); % coordinate
   DNurbsbasis = DNurbsLeibnitz(npa,knota,wta,t);
```

```
  Su0(1:nba-npa,i)=DNurbsbasis(1:nba-npa,2*npa+2);%non-derivative
  Su1(1:nba-npa,i)=DNurbsbasis(1:nba-npa,2*npa+2+1);%1st-derivative
end
for i=1:ngp
  t = Gpw(i,1); % coordinate
  DNurbsbasis = DNurbsLeibnitz(npb,knotb,wtb,t);
  Sv0(1:nbb-npb,i)=DNurbsbasis(1:nbb-npb,2*npb+2); %non-derivative
  Sv1(1:nbb-npb,i)=DNurbsbasis(1:nbb-npb,2*npb+2+1);%1st-derivative
end
for i=1:ngp
  t = Gpw(i,1); % coordinate
  DNurbsbasis = DNurbsLeibnitz(npc,knotc,wtc,t);
  St0(1:nbc-npc,i)=DNurbsbasis(1:nbc-npc,2*npc+2); %non-derivative
  St1(1:nbc-npc,i)=DNurbsbasis(1:nbc-npc,2*npc+2+1);%1st-derivative
end
%K11
for i=1:ngp
  for j=1:nba-npa
    for k=1:nba-npa
      Kmtx(j,k) = Kmtx(j,k)+E*A*Su1(j,i)/Jac*Su1(k,i)/Jac*Jac*Gpw
(i,2);
    end
  end
end
%K22
nna = nba-npa; nnb = nbb-npb;
for i=1:ngp
  for j=1:nnb
    for k=1:nnb
      Kmtx(j+nna,k+nna) = Kmtx(j+nna,k+nna)...
        +kp*G*A*Sv1(j,i)/Jac*Sv1(k,i)/Jac*Jac*Gpw(i,2);
    end
  end
end
%K23
nnc = nbc-npc;
for i=1:ngp
  for j=1:nnb
    for k=1:nnc
      Kmtx(j+nna,k+nna+nnb) = Kmtx(j+nna,k+nna+nnb)...
        -kp*G*A*Sv1(j,i)/Jac*St0(k,i)*Jac*Gpw(i,2);
    end
  end
end
%K32
```

```
for i=1:ngp
  for j=1:nnc
    for k=1:nnb
      Kmtx(j+nna+nnb,k+nna) = Kmtx(j+nna+nnb,k+nna)...
          -kp*G*A*St0(j,i)*Sv1(k,i)/Jac*Jac*Gpw(i,2);
    end
  end
end
% K33
for i=1:ngp
  for j=1:nnc
    for k=1:nnc
        Kmtx(j+nna+nnb,k+nna+nnb) = Kmtx(j+nna+nnb,k+nna+nnb)...
            +kp*G*A*St0(j,i)*St0(k,i)*Jac*Gpw(i,2)...
            +E*I*St1(j,i)/Jac*St1(k,i)/Jac*Jac*Gpw(i,2);
    end
  end
end
% disp('The Stiffness Matrix');
% disp(Kmtx);
% Mass matrix K
% M11
for i=1:ngp
  for j=1:nna
    for k=1:nna
        Mmtx(j,k) = Mmtx(j,k)+rho*A*Su0(j,i)*Su0(k,i)*Jac*Gpw(i,2);
    end
  end
end
% M22
for i=1:ngp
  for j=1:nnb
    for k=1:nnb
        Mmtx(j+nna,k+nna) = Mmtx(j+nna,k+nna)...
            +rho*A*Sv0(j,i)*Sv0(k,i)*Jac*Gpw(i,2);
    end
  end
end
% M33
for i=1:ngp
  for j=1:nnc
    for k=1:nnc
        Mmtx(j+nna+nnb,k+nna+nnb) = Mmtx(j+nna+nnb,k+nna+nnb)...
            +rho*I*St0(j,i)*St0(k,i)*Jac*Gpw(i,2);
    end
  end
end
```

```
% disp('The Mass Matrix');
% disp(Mmtx);
% Loading Vector
Lvec=[0 0 0 0 0 -500 0 0 0]';
% Condensation
i=4; j=5; k=9;
[KMC,MMC,FVC] = Condensation(Kmtx,Mmtx,Lvec,i,j,k); % Condensing v
i=6; j=6; k=7;
[KMC,MMC,FVC] = Condensation(KMC,MMC,FVC,i,j,k); % Condensing theta
% Swapping the matrices
KMC=KMC([1,3,5,2,4,6],:); % row
KMC=KMC(:,[1,3,5,2,4,6]); % column
MMC=MMC([1,3,5,2,4,6],:); % row
MMC=MMC(:,[1,3,5,2,4,6]); % column
FVC=FVC([1,3,5,2,4,6],:); % row
disp('The Stiffness Matrix after condensation');
disp(KMC);
disp('The Mass Matrix after condensation');
disp(MMC);
```

5.5.2 Condensation Function List

```
% Condensation.m
function [KMC,MMC,FVC] = Condensation(KM,MM,FV,nA,nB,nC)
n1 = nA - 1;
n2 = nB + 1;
% Extracting KM
K11 = KM(1:n1,1:n1); K12 = KM(1:n1,nA:nB); K13 = KM(1:n1,n2:nC);
K21 = KM(nA:nB,1:n1); K22 = KM(nA:nB,nA:nB); K23 = KM(nA:nB,n2:nC);
K31 = KM(n2:nC,1:n1); K32 = KM(n2:nC,nA:nB); K33 = KM(n2:nC,n2:nC);
% Extracting MM
M11 = MM(1:n1,1:n1); M12 = MM(1:n1,nA:nB); M13 = MM(1:n1,n2:nC);
M21 = MM(nA:nB,1:n1); M22 = MM(nA:nB,nA:nB); M23 = MM(nA:nB,n2:nC);
M31 = MM(n2:nC,1:n1); M32 = MM(n2:nC,nA:nB); M33 = MM(n2:nC,n2:nC);
% Extracting FV
F1 = FV(1:n1); F2 = FV(nA:nB); F3 = FV(n2:nC);
% Condensation
K22I =inv(K22);
%
M21B = M21 - M22 * K22I * K21;
M23B = M23 - M22 * K22I * K23;
K11B = K11 - K12 * K22I * K21;
```

```
M11BB = M11 - M12 * K22I * K21 - K12 * K22I * M21B;
F1B = F1 - K12 * K22I * F2;
K13B = K13 - K12 * K22I * K23;
M13BB = M13 - M12 * K22I * K23 - K12 * K22I * M23B;
K31B = K31 - K32 * K22I * K21;
M31BB = M31 - M32 * K22I * K21 - K32 * K22I * M21B;
K33B = K33 - K32 * K22I * K23;
M33BB = M33 - M32 * K22I * K23 - K32 * K22I * M23B;
F3B = F3 - K32 * K22I * F2;
%
KMC = [K11B K13B; K31B K33B];
MMC = [M11BB M13BB; M31BB M33BB];
FVC = [ F1B; F3B];
Return
```

5.5.3 *KMmatrixCircTiNURBSCondensed Program List*

MATLAB code KMmatrixCircTiNURBSCondensed.m and functions Condensation.m (Sect. 5.5.2), Legendre.m (Sect. 2.3.2), and DNurbsLeibnitz.m (Sect. 1.3.14) integrate the stiffness and mass matrices of a circular curved Timoshenko beam numerically by using NURBS functions. As an example, the stiffness and mass matrices of the circular curved Timoshenko beam shown in Fig. 4.10 are reduced to a two-node six degree of freedom beam element which is shown in Fig. 5.4.

We can see from Fig. 5.4 that before the condensation, the stiffness and mass matrices had twelve degree of freedom beam element (Fig. 4.11), since from the horizontal displacement, u, v, and rotational θ were assumed to be third order of NURBS. The stiffness and mass matrices of the element after condensed and

The Stiffness Matrix after condensation

112.9908	-56.5892	4.0381	-56.7742	-112.9676	40.9352
-56.5892	112.9936	40.9896	112.9676	56.3794	4.1131
4.0381	40.9896	39.0013	40.9352	-4.1131	-3.0226
-56.7742	112.9676	40.9352	112.9908	56.5892	4.0381
-112.9676	56.3794	-4.1131	56.5892	112.9936	-40.9896
40.9352	4.1131	-3.0226	4.0381	-40.9896	39.0013

The Mass Matrix after condensation

76.4178	1.7272	2.1397	-4.9576	22.2854	-6.9527
1.7272	37.0197	3.2151	-22.2854	-8.3183	-1.0567
2.1397	3.2151	2.9535	-6.9527	1.0567	-0.1793
-4.9576	-22.2854	-6.9527	76.4178	-1.7272	2.1397
22.2854	-8.3183	1.0567	-1.7272	37.0197	-3.2151
-6.9527	-1.0567	-0.1793	2.1397	-3.2151	2.9535

Fig. 5.4 The stiffness and mass matrices of a circular curved Timoshenko beam element example

rearranged are exactly the same with the classical two-node beam element with six degrees of freedom.

```
% KMmatrixCircTiNURBSCondensed.m
% Construct matrices K and M
% Based on the Timoshenko Beam Theory by using NURBS shape
functions
clear all;clc;
E = 1525; A = 0.2; I = 0.014; kp = 0.85; nu = 0.3;
rho = 630; G = E/2/(1+nu); R = 0.8; alp1 = 3*pi/4; alp2 = pi/4;
% NURBS for u displacement data
npa = 3;
knota = [-ones(1,npa+1) ones(1,npa+1)]; knota = [knota ones(1,
npa)];
nba = size(knota,2)-npa-1;
wta = ones(nba-npa);
% NURBS for v displacement data
npb = 3;
knotb = [-ones(1,npb+1) ones(1,npb+1)]; knotb = [knotb ones(1,
npb)];
nbb = size(knotb,2)-npb-1;
wtb = ones(nbb-npb);
% NURBS for th displacement data
npc = 3;
knotc = [-ones(1,npc+1) ones(1,npc+1)]; knotc = [knotc ones(1,
npc)];
nbc = size(knotc,2)-npc-1;
wtc = ones(nbc-npc);
% Jacobi and Gauss
Jac = abs(alp2-alp1)/2*R; % Constant for a circular arc
ngp = 4; Gpw = Legendre(ngp);
dim = npa+1+npb+1+npc+1;
Kmtx = zeros(dim,dim); Mmtx = zeros(dim,dim);
Su0 = zeros(nba-npa,ngp); Sv0 = zeros(nbb-npb,ngp);
St0 = zeros(nbc-npc,ngp); Su1 = zeros(nba-npa,ngp);
Sv1 = zeros(nbb-npb,ngp); St1 = zeros(nbc-npc,ngp);
for i=1:ngp
 t = Gpw(i,1); % coordinate
 DNurbsbasis = DNurbsLeibnitz(npa,knota,wta,t);
  Su0(1:nba-npa,i)=DNurbsbasis(1:nba-npa,2*npa+2+0); % non-deri-
vative
  Su1(1:nba-npa,i)=DNurbsbasis(1:nba-npa,2*npa+2+1); % 1st-deri-
vative
end
```

```
for i=1:ngp
 t = Gpw(i,1); % coordinate
 DNurbsbasis = DNurbsLeibnitz(npb,knotb,wtb,t);
  Sv0(1:nbb-npb,i)=DNurbsbasis(1:nbb-npb,2*npb+2+0); % non-deri-
vative
  Sv1(1:nbb-npb,i)=DNurbsbasis(1:nbb-npb,2*npb+2+1); % 1st-deri-
vative
end
for i=1:ngp
 t = Gpw(i,1); % coordinate
 DNurbsbasis = DNurbsLeibnitz(npc,knotc,wtc,t);
  St0(1:nbc-npc,i)=DNurbsbasis(1:nbc-npc,2*npc+2+0); % non-deri-
vative
  St1(1:nbc-npc,i)=DNurbsbasis(1:nbc-npc,2*npc+2+1); % 1st-deri-
vative
end
% K11
nna = nba-npa;
for i=1:ngp
   for j=1:nna
     for k=1:nna
         Kmtx(j,k)  =  Kmtx(j,k)+E*A*Su1(j,i)/Jac*Su1(k,i)/Jac*-
         Jac*Gpw(i,2)...
        +kp*G*A*Su0(j,i)/R*Su0(k,i)/R*Jac*Gpw(i,2);
     end
   end
end
% K12
nnb = nbb-npb;
for i=1:ngp
   for j=1:nna
     for k=1:nnb
        Kmtx(j,k+nna) = Kmtx(j,k+nna)...
              +E*A*Su1(j,i)/Jac*Sv0(k,i)/R*Jac*Gpw(i,2)...
              -kp*G*A*Su0(j,i)/R*Sv1(k,i)/Jac*Jac*Gpw(i,2);
        Kmtx(k+nna,j) = Kmtx(j,k+nna); % Symmetric K21
     end
   end
end
% K13
nnc = nbc-npc;
for i=1:ngp
 for j=1:nna
 for k=1:nnc
 Kmtx(j,k+nna+nnb) = Kmtx(j,k+nna+nnb)...
```

```
 +kp*G*A*Su0(j,i)/R*St0(k,i)*Jac*Gpw(i,2);
 Kmtx(k+nna+nnb,j) = Kmtx(j,k+nna+nnb); % Symmetric K31
 end
 end
end
% K22
for i=1:ngp
 for j=1:nnb
 for k=1:nnb
 Kmtx(j+nna,k+nna) = Kmtx(j+nna,k+nna)...
 +E*A*Sv0(j,i)/R*Sv0(k,i)/R*Jac*Gpw(i,2)...
 +kp*G*A*Sv1(j,i)/Jac*Sv1(k,i)/Jac*Jac*Gpw(i,2);
 end
 end
end
% K23
for i=1:ngp
 for j=1:nnb
 for k=1:nnc
 Kmtx(j+nna,k+nna+nnb) = Kmtx(j+nna,k+nna+nnb)...
 -kp*G*A*Sv1(j,i)/Jac*St0(k,i)*Jac*Gpw(i,2);
 Kmtx(k+nna+nnb,j+nna) = Kmtx(j+nna,k+nna+nnb); % Symmetric K32
 end
 end
end
% K33
for i=1:ngp
 for j=1:nnc
 for k=1:nnc
 Kmtx(j+nna+nnb,k+nna+nnb) = Kmtx(j+nna+nnb,k+nna+nnb)...
 +kp*G*A*St0(j,i)*St0(k,i)*Jac*Gpw(i,2)...
 +E*I*St1(j,i)/Jac*St1(k,i)/Jac*Jac*Gpw(i,2);
 end
 end
end
disp('The Stiffness Matrix');
disp(Kmtx);
% Mass matrix K
% M11
for i=1:ngp
 for j=1:nna
 for k=1:nna
 Mmtx(j,k) = Mmtx(j,k)+rho*A*Su0(j,i)*Su0(k,i)*Jac*Gpw(i,2);
 end
 end
```

```
end
% M22
for i=1:ngp
 for j=1:nnb
 for k=1:nnb
 Mmtx(j+nna,k+nna) = Mmtx(j+nna,k+nna)...
 +rho*A*Sv0(j,i)*Sv0(k,i)*Jac*Gpw(i,2);
 end
 end
end
% M33
for i=1:ngp
 for j=1:nnc
 for k=1:nnc
 Mmtx(j+nna+nnb,k+nna+nnb) = Mmtx(j+nna+nnb,k+nna+nnb)...
 +rho*I*St0(j,i)*St0(k,i)*Jac*Gpw(i,2);
 end
 end
end
disp('The Mass Matrix');
disp(Mmtx);
% Loading Vector
Lvec=[0 0 0 0 0 0 0 0 0 0 0 -500]';
% Condensation
i=2; j=3; k=12;
[KMC,MMC,FVC] = Condensation(Kmtx,Mmtx,Lvec,i,j,k); % Condensing
u
i=4; j=5; k=10;
[KMC,MMC,FVC] = Condensation(KMC,MMC,FVC,i,j,k); % Condensing v
i=6; j=7; k=8;
[KMC,MMC,FVC] = Condensation(KMC,MMC,FVC,i,j,k); % Condensing
theta
% Swapping the matrices
KMC=KMC([1,3,5,2,4,6],:); % row
KMC=KMC(:,[1,3,5,2,4,6]); % column
MMC=MMC([1,3,5,2,4,6],:); % row
MMC=MMC(:,[1,3,5,2,4,6]); % column
FVC=FVC([1,3,5,2,4,6],:); % row
disp('The Stiffness Matrix after condensation');
disp(KMC);
disp('The Mass Matrix after condensation');
disp(MMC);
```

Further Reading

Bathe KJ (1982) Finite element procedures in engineering analysis. Prentice-Hall, Englewood Cliffs

Cook RD (1981) Concepts and applications of finite element analysis, 2nd edn. Wiley, New York

Hart GC, Wong K (2000) Structural dynamics for structural engineers. Wiley, New York

Chapter 6
Straight Beam Element Examples

6.1 Euler-Bernoulli Beam

In the Euler-Bernoulli straight beam theory, the plane sections remain plane during the deformation (Sects. 3.4.1, 3.5.1, and 3.6). The program lists in Sect. 6.2 can be adopted into the general finite element software to solve the static and free vibration problems.

It should be noted that all the program lists given in the following chapters are solved by using the least element formulation. The standard procedure such as assembling technique, coordinate transformation, and sub-element division method in general finite element program is omitted from the lists because it will take too much space in this book.

6.1.1 Static Problem

Consider a cantilever beam shown in Fig. 6.1. The beam has geometry and property depicted in the figure. The degrees of polynomial used in the NURBS functions are $p_u = 1$ and $p_v = 3$.

The beam is subjected to various loading conditions. One element with four Gauss integration points is used in the analysis. The results of static analyses are given in Table 6.1.

Next, consider a simply supported beam shown in Fig. 6.2. The beam has geometry and property depicted in the figure.

The beam is subjected to several loading conditions. One element with four Gauss integration points is used in the analysis. The results of static analyses are given in Table 6.2.

Results which are listed in Tables 6.1 and 6.2 show excellent agreement with those of the classical beam theory equations.

B.S. Gan, *An Isogeometric Approach to Beam Structures*,
DOI 10.1007/978-3-319-56493-7_6

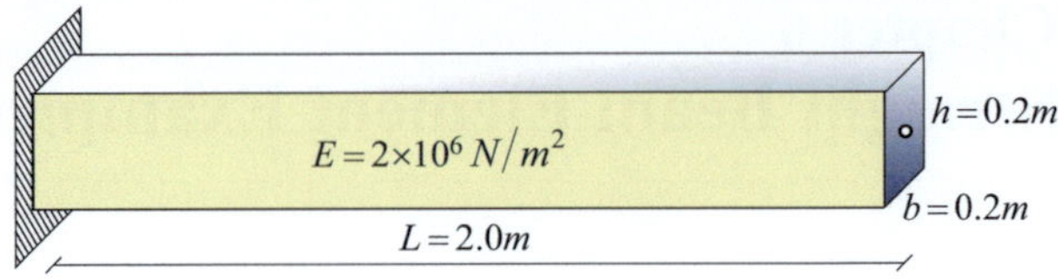

Fig. 6.1 Cantilever Euler-Bernoulli straight beam under various loading examples

Table 6.1 Results of static analyses of cantilever beam subjected to various loadings

	Deflection at free end (10^{-3} m)		Rotation at free end (10^{-3} rad)	
Type of loading	Analysis	Theory	Analysis	Theory
Concentrated load P_y at free end (1 N)	−10.00	$v_{max} = -\frac{PL^3}{3EI} = -10.00$	−7.500	$\theta_{max} = -\frac{PL^2}{2EI} = -7.500$
Bending moment M_z at free end (1 N.m)	7.500	$v_{max} = \frac{ML^4}{2EI} = 7.500$	7.500	$\theta_{max} = \frac{ML}{EI} = 7.500$
Distributed load q_y along the span (1 N/m)	7.500	$v_{max} = \frac{qL^4}{8EI} = 7.500$	5.000	$\theta_{max} = \frac{qL^3}{6EI} = 5.000$

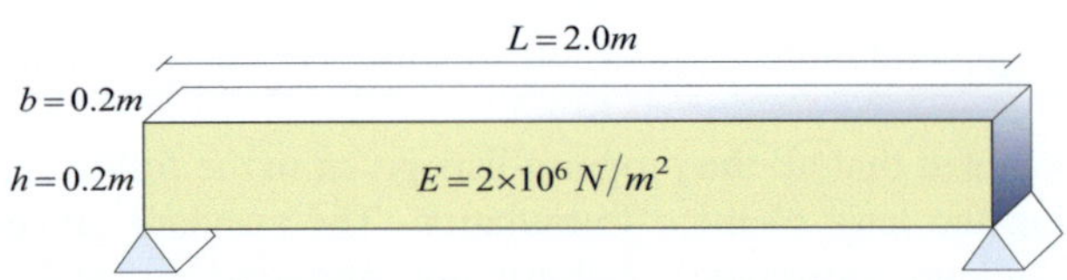

Fig. 6.2 Simply supported Euler-Bernoulli straight beam under various loading examples

Table 6.2 Results of static analyses of simply supported beam subjected to various loadings

	Deflection at mid-span (10^{-3} m)		Rotation at supports (10^{-3} rad)	
Type of loading	Analysis	Theory	Analysis	Theory
Concentrated load P_y at mid-span (1 N)	−0.6250	$v_{max} = -\frac{PL^3}{48EI} = -0.6250$	±0.9375	$\theta_{max} = \pm\frac{PL^2}{16EI} = \pm 0.9375$
Bending moment M_z at mid-span (1 N.m)	0.0	0.0	−0.3125	$\theta_{max} = -\frac{ML}{24EI} = -0.3125$
Distributed load q_y along the span (1 N/m)	−0.78125	$v_{max} = -\frac{5qL^4}{384EI} = -0.78125$	±1.250	$\theta_{max} = \pm\frac{qL^3}{24EI} = \pm 1.250$

6.1.2 Free Vibration Problem

Consider the simply supported beam shown in Fig. 6.2 as a free vibration problem. We need to divide the beam element into sub-elements to get several mode shapes

Table 6.3 Free vibration results of the first six normalized natural frequencies of simply supported Timoshenko beam

Mode number i	ω_{EB-n}	$\frac{\omega_i}{\omega_{EB-n}}$ 2 Elements	3 Elements	4 Elements
1	20.1462	1.0039	1.0008	1.0003
2	80.5850	1.1099	1.0118	1.0039
3	181.3162	1.2399	1.1099	1.0183
4	322.3399	1.2716	1.1609	1.1099
5	503.6561	–	1.3294	1.1291
6	725.2648	–	1.2716	1.2399

and frequencies (ω) of the vibration. The next example analyzes the natural frequencies of the Euler-Bernoulli beam.

The beam with the length $L = 10$ m, the rectangular cross section with width $b = 1.5$ m, and depth $h = 2$ m. The elastic modulus $E = 1$ N/m^2, and the density $\rho = 1$ kg/m^3. Four Gauss integration points are necessary to integrate the mass matrix. The degrees of polynomial used in the NURBS functions are $p_u = 1$ and $p_v = 3$.

The free vibration problems which were analyzed by using two-, three-, and four-element sub-division are given in Table 6.3. The results are compared with the exact values ($\omega_{EB\text{-}n}$) from the nth mode of free vibration of Euler-Bernoulli beam theory which is defined by

$$\omega_{EB-n} = \left(\frac{n\ \pi}{L}\right)^2 \sqrt{\frac{EI}{\rho A}} \tag{6.1}$$

From Table 6.3, we can observe that by increasing the number of element to four sub-divisions, we can get better results with the first six frequency modes of the Euler-Bernoulli straight beam free vibration problems.

6.2 Program Lists for Solving Straight Euler-Bernoulli Beams

MATLAB code StraightEuBeBeam.m and functions Condensation.m (Sect. 5.5.2), Legendre.m (Sect. 2.3.2), KMEuBeStraight.m (Sect. 6.2.2), and DNurbsLeibnitz.m (Sect. 1.3.14) create element stiffness and mass matrices of a straight Euler-Bernoulli beam numerically by using NURBS functions. The stiffness and mass matrices are reduced to a two-node six-degree-of-freedom beam element and solved for static and free vibration analyses. The results are presented in Table 6.3.

6.2.1 StraightEuBeBeam Program List

```
% StraightEuBeBeam.m
% Solving Straight Euler-Bernoulli Beam Problems
% Euler-Bernoulli Beam Theory by using NURBS shape functions
clear all;clc;
%ne = 1;                    % Element Division Table 6.1
%ne = 2;                    % Element Division Table 6.2
ne = 4;                     % Element Division Table 6.3
% Material Data
Data(1) = 2e6;              % Young's Modulus E
Data(2) = 2.0/ne;           % Length Le
Data(3) = 0.2*0.2;          % Cross Section Area A
Data(4) = 0.2*0.2^3/12;     % Second Moment of Inertia I
Data(5) = 100;              % Density rho
Data(6) = 1;                % Order Polynomial Degree of U npa
Data(7) = 3;                % Order Polynomial Degree of V npb
Data(8) = 4;                % Gauss Points ngp
% Jacobi and Gauss data
L = Data(2);
Jac = L/2;
ngp = Data(8);
Gpw = Legendre(ngp);
Bmtx = zeros(1,4);
% Global Matrices
KV = zeros(6+3*(ne-1),6+3*(ne-1));
MV = zeros(6+3*(ne-1),6+3*(ne-1));
FV = zeros(6+3*(ne-1),1);
for ii = 1:ne
    % Constructing Element Matrices
    [KMC,MMC,Sv0] = KMEuBeStraight(Data,Jac,Gpw);
    Fload=zeros(6,1);
    for i=1:ngp
      Bmtx(1) = Sv0(1,i) + Sv0(2,i);
      Bmtx(2) = Sv0(3,i) + Sv0(4,i);
      Bmtx(3) = L/3*Sv0(2,i);
      Bmtx(4) =-L/3*Sv0(3,i);
      % Distributed loading qy = -1.0 N/m (Loading 3)
      for j=1:4
        %Fload(2+j,1)= Fload(2+j,1)+ 1.0*Bmtx(j)*Jac*Gpw(i,2);
      end
      % Table 6.1 & 6.2 ************************************
```

```
    end
    Fload=Fload([1,3,5,2,4,6],:); % row swapping
    % Assembling matrices
    if ii==1 % only the first element
      KV(1:6,1:6)=KMC(:,:);
      MV(1:6,1:6)=MMC(:,:);
      FV(1:6,1)=Fload(:,:);
    else
      n1=1+3*(ii-1);
      n2=6+3*(ii-1);
      KV(n1:n2,n1:n2)=KV(n1:n2,n1:n2)+KMC(:,:);
      MV(n1:n2,n1:n2)=MV(n1:n2,n1:n2)+MMC(:,:);
      FV(n1:n2,1)=FV(n1:n2,1)+Fload(:,:);
    end
end % end of ne element sub division ii
% Eliminating fixed DOFs for cantilever support Table 6.1 (ne=1)
% KV=KV([4:end],:); % extracting free DOF rows
% KV=KV(:,[4:end]); % extracting free DOF columns
% MV=MV([4:end],:); % extracting free DOF rows
% MV=MV(:,[4:end]); % extracting free DOF columns
% FV=FV([4:end],:); % extracting free DOF rows
% Loading at free end of the beam
%FV = [0; -1; 0]; % Py = 1 N (Loading 1)
%FV = [0; 0; 1]; % Mz = 1 N (Loading 2)
% Table 6.1 **************************************************
% Eliminating fixed DOFs for simply supported Table 6.2 (ne=2)
% KV=KV([3:end-3,end],:); % extracting free DOF rows
% KV=KV(:,[3:end-3,end]); % extracting free DOF columns
% MV=MV([3:end-3,end],:); % extracting free DOF rows
% MV=MV(:,[3:end-3,end]); % extracting free DOF columns
% FV=FV([3:end-3,end],:); % extracting free DOF rows
% Loading at mid-span of the beam
%FV = [0; 0; -1; 0; 0]; % Py = 1 N (Loading 1)
%FV = [0; 0; 0; 1; 0]; % Mz = 1 N (Loading 2)
% Table 6.2 **************************************************
% Static Analysis Table 6.1 & Table 6.2 ************************
% disp('Loading Vector - Static Analysis');
% disp(FV);
% disp('Displacement Vector - Static Analysis');
% Uvec=KV\FV;
% disp(Uvec(:)*1000);
% Static Analysis Table 6.1 & Table 6.2 ************************
% Free Vibration Analysis Table 6.3 ****************************
% Eliminating fixed DOFs for simply supported Table 6.3 (ne=2,3,4)
% KV=KV([3,5,6,9],:); % extracting free DOF rows 2 elem
```

```
% KV=KV(:,[3,5,6,9]); % extracting free DOF columns
% MV=MV([3,5,6,9],:); % extracting free DOF rows
% MV=MV(:,[3,5,6,9]); % extracting free DOF columns 2 elem
% KV=KV([3,5,6,8,9,12],:); % extracting free DOF rows 3 elem
% KV=KV(:,[3,5,6,8,9,12]); % extracting free DOF columns
% MV=MV([3,5,6,8,9,12],:); % extracting free DOF rows
% MV=MV(:,[3,5,6,8,9,12]); % extracting free DOF columns 3 elem
KV=KV([3,5,6,8,9,11,12,15],:); % extracting free DOF rows 4 elem
KV=KV(:,[3,5,6,8,9,11,12,15]); % extracting free DOF columns
MV=MV([3,5,6,8,9,11,12,15],:); % extracting free DOF rows
MV=MV(:,[3,5,6,8,9,11,12,15]); % extracting free DOF columns
4 elem
% Free Vibration Analysis (Eigenvalue Analysis)
[vec, lam2] = eig(KV,MV);
% Normalized Natural Frequency
lambda=sort(diag(sqrt(lam2)));
ni = size(KV,1);
if ni>6
   ni = 6;
end
omegaEB=zeros(ni,1); omegaN=zeros(ni,1);
for i=1:ni
   omegaN(i,1)=((i*pi/Data(2)/ne)^2*...
    sqrt(Data(1)*Data(4)/Data(5)/Data(3)));
 omegaEB(i,1) = lambda(i)/omegaN(i,1);
end
disp('First Six Normalized Natural Frequencies');
disp(omegaN(1:ni,1));
disp(omegaEB(1:ni,1));
% Free Vibration Analysis Table 6.3 ***************************
```

6.2.2 KMEuBeStraight Function List

```
function [KMC,MMC,Sv0] = KMEuBeStraight(Data,Jac,Gpw)
% KMEuBeStraight.m
% Constructing Stiffness and Mass Matrices of a Straight Beam
Element
% Followed by condensation and swapping to the standard DOFs
% Euler-Bernoulli Beam Theory by using NURBS shape functions
E = Data(1);              % Young's Modulus E
L = Data(2);              % % Length Le
A = Data(3);              % Cross Section Area A
```

```
I = Data(4);            % Second Moment of Inertia I
rho = Data(5);          % Density rho
npa = Data(6);          % Order Polynomial Degree of U npa
npb = Data(7);          % Order Polynomial Degree of V npb
ngp = Data(8);          % Gauss Points ngp
dim = npa+1+npb+1;
KMC = zeros(dim,dim);
MMC = zeros(dim,dim);
% NURBS for u displacement data
knota = [-ones(1,npa+1) ones(1,npa+1)]; knota = [knota ones(1,
npa)];
nba = size(knota,2)-npa-1;
wta = ones(nba-npa);
nna = nba-npa;
% NURBS for v and theta displacement data
knotb = [-ones(1,npb+1) ones(1,npb+1)]; knotb = [knotb ones(1,
npb)];
nbb = size(knotb,2)-npb-1;
wtb = ones(nbb-npb);
nnb = nbb-npb;
%
Su0 = zeros(nba-npa,ngp); Su1 = zeros(nba-npa,ngp);
Sv0 = zeros(nbb-npb,ngp);
St0 = zeros(nbb-npb,ngp); Sv2 = zeros(nbb-npb,ngp);
for i=1:ngp
    t = Gpw(i,1); % coordinate
    DNurbsbasis = DNurbsLeibnitz(npa,knota,wta,t);
    Su0(1:nba-npa,i)=DNurbsbasis(1:nba-npa,npa+5); % non-derivative
   Su1(1:nba-npa,i)=DNurbsbasis(1:nba-npa,npa+7); % 1st-derivative
end
for i=1:ngp
   t = Gpw(i,1); % coordinate
   DNurbsbasis = DNurbsLeibnitz(npb,knotb,wtb,t);
   Sv0(1:nbb-npb,i)=DNurbsbasis(1:nbb-npb,npb+5); % non-derivative
   Sv1(1:nbb-npb,i)=DNurbsbasis(1:nbb-npb,npb+7); % 1st-derivative
   Sv2(1:nbb-npb,i)=DNurbsbasis(1:nbb-npb,npb+8); % 2nd-derivative
end
% K11
for i=1:ngp
   for j=1:nba-npa
     for k=1:nba-npa
        KMC(j,k) = KMC(j,k)+E*A*Su1(j,i)/Jac*Su1(k,i)/Jac*Jac*Gpw
        (i,2);
     end
   end
```

```
end
% K22
for i=1:ngp
  Bmtx(1) = Sv2(1,i)/Jac/Jac + Sv2(2,i)/Jac/Jac;
  Bmtx(2) = Sv2(3,i)/Jac/Jac + Sv2(4,i)/Jac/Jac;
  Bmtx(3) = L/3*Sv2(2,i)/Jac/Jac;
  Bmtx(4) =-L/3*Sv2(3,i)/Jac/Jac;
 for j=1:4
    for k=1:4
        KMC(j+nna,k+nna) = KMC(j+nna,k+nna)...
        +E*I*Bmtx(j)*Bmtx(k)*Jac*Gpw(i,2);
    end
  end
end
% Mass matrix K
% M11 translation mass rho*A
for i=1:ngp
    for j=1:nna
      for k=1:nna
          MMC(j,k) = MMC(j,k)+rho*A*Su0(j,i)*Su0(k,i)*Jac*Gpw(i,2);
      end
    end
end
% M22A translation mass rho*A
for i=1:ngp
  Bmtx(1) = Sv0(1,i) + Sv0(2,i);
  Bmtx(2) = Sv0(3,i) + Sv0(4,i);
  Bmtx(3) = L/3*Sv0(2,i);
  Bmtx(4) =-L/3*Sv0(3,i);
  for j=1:4
      for k=1:4
        MMC(j+nna,k+nna) = MMC(j+nna,k+nna)...
        +rho*A*Bmtx(j)*Bmtx(k)*Jac*Gpw(i,2);
      end
  end
end
% Swapping the matrices
KMC=KMC([1,3,5,2,4,6],:); % row
KMC=KMC(:,[1,3,5,2,4,6]); % column
MMC=MMC([1,3,5,2,4,6],:); % row
MMC=MMC(:,[1,3,5,2,4,6]); % column
return
```

6.3 Timoshenko Beam

In the Timoshenko straight beam theory, the plane of the cross section is considered not to remain perpendicular to the beam axis during the deformation (Sects. 3.4.2, 3.5.2, and 3.7). The shear deformation is taken into account in the beam formulation.

The program lists in Sect. 6.4 can be adopted into the general finite element software to solve the static and free vibration problems.

6.3.1 Static Problem

Consider a cantilever beam shown in Fig. 6.1. The beam has geometry and property depicted in the figure. To consider the effect of shear deformation in the beam, the Poisson's ratio $\nu = 0.3$ and the shear correction coefficient $\kappa = 5/6$ are used.

The beam is subjected to several loading conditions. One element with four Gauss integration points is used in the analysis. The degrees of polynomial used in the NURBS functions are $p_u = 1$, $p_v = 3$, and $p_\theta = 2$. The four Gauss integration points are necessary to integrate the mass matrix. The results of static analyses are given in Table 6.4.

Next, consider a simply supported beam shown in Fig. 6.2. The beam has geometry and property depicted in the figure. To consider the effect of shear deformation in the beam, the Poisson's ratio $\nu = 0.3$ and the shear correction coefficient $\kappa = 5/6$ are used. The beam is subjected to several loading conditions. Two elements with four Gauss integration points are used in the analysis. The degrees of polynomial used in the NURBS functions are $p_u = 1, p_v = 3$, and $p_\theta = 2$. The four Gauss integration points are necessary to integrate the mass matrix. The results of static analyses are given in Table 6.5.

Table 6.4 Results of static analyses of cantilever beam subjected to various loadings

	Deflection at free end (10^{-3} m)		Rotation at free end (10^{-3} rad)	
Type of loading	Analysis	Theory	Analysis	Theory
Concentrated load P_y at free end (1 N)	−10.078	$v_{\max} = -\frac{PL^3}{3EI} - \frac{PL}{\kappa GA} = -10.078$	−7.500	$\theta_{\max} = -\frac{PL^2}{2EI} = -7.500$
Bending moment M_z at free end (1 N.m)	7.500	$v_{\max} = \frac{ML^4}{2EI} = 7.500$	7.500	$\theta_{\max} = \frac{ML}{EI} = 7.500$
Distributed load q_y along the span (1 N/m)	−7.578	$v_{\max} = -\frac{qL^4}{8EI} - \frac{qL^2}{2\kappa GA} = -7.578$	−5.000	$\theta_{\max} = -\frac{qL^3}{6EI} = -5.000$

Table 6.5 Results of static analyses of simply supported beam subjected to various loadings

	Deflection at mid-span (10^{-3} m)		Rotation at supports (10^{-3} rad)	
Type of loading	Analysis	Theory	Analysis	Theory
Concentrated load P_y at mid-span (1 N)	−0.6445	$v_{\max}= -\frac{PL^3}{48EI} - \frac{PL}{4\kappa GA} = -0.6445$	±0.9375	$\theta_{\max}= \frac{PL^2}{16EI} = \pm 0.9375$
Bending moment M_z at mid-span (1 N.m)	0.0	0.0	−0.2930	$\theta_{\max}= -\frac{ML}{24EI} - \frac{M}{\kappa GAL} = -0.2930$
Distributed load q_y along the span (1 N/m)	−0.80075	$v_{\max}= -\frac{5qL^4}{384EI} - \frac{qL^2}{8\kappa GA} = -0.80075$	±1.250	$\theta_{\max}= \frac{qL^3}{24EI} = \pm 1.250$

Table 6.6 Free vibration results of the first six normalized natural frequencies of simply supported Timoshenko beam

Mode number		$\frac{\omega_i}{\omega_{TI-n}}$		
i	ω_{TI-n}	2 Elements	3 Elements	4 Elements
1	5.1163	1.0062	1.0019	1.0009
2	19.5514	1.1620	1.0254	1.0122
3	41.1912	1.3573	1.2167	1.0501
4	67.8131	1.5429	1.3131	1.2796
5	97.6715		1.6388	1.3207
6	129.5700		1.7336	1.5417

6.3.2 *Free Vibration Problem*

Consider the simply supported beam shown in Fig. 6.2 as a free vibration prob-lem. We need to divide the beam element into sub-elements to get several mode shapes and frequencies (ω) of the vibration. The next example analyses the natural frequencies of the Euler-Bernoulli beam.

The beam with the length $L = 2$ m, the rectangular cross section with width $b = 0.2$ m and depth $h = 0.2$ m. The elastic modulus $E = 2.0$ MN/m^2, the Poisson's ratio $\nu = 0.3$, the shear correction coefficient $\kappa = 5/6$, and the density $\rho = 1500$ kg/m^3. Four Gauss integration points are necessary to integrate the mass matrix. The degrees of polynomial used in the NURBS functions are $p_u = 1, p_v = 3$, and $p_\theta = 2$.

The free vibration problems which were analyzed by using two-, three-, and four-element sub-division are given in Table 6.6. The results are compared with the exact values ($\omega_{TI\text{-}n}$) from the nth mode of free vibration of Timoshenko beam theory which is obtained from Cazzani et al. (2016) and Lee (2009) by using the Spectral Element Analysis.

From Table 6.6, we can observe that by further increasing the number of sub-division element, we can get convergence results with the first six frequency modes of the Timoshenko straight beam free vibration problems.

6.4 Program Lists for Solving Straight Timoshenko Beams

MATLAB code StraightTimoBeam.m and functions Condensation.m (Sect. 5.5.2), Legendre.m (Sect. 2.3.2), KMTimoStraight.m (Sect. 6.4.2), and DNurbsLeibnitz.m (Sect. 1.3.14) create element stiffness and mass matrices of a straight Timoshenko beam numerically by using NURBS functions. The stiffness and mass matrices are reduced to a two-node six-degree-of-freedom beam element and solved for static and free vibration analyses. The results are presented in Tables 6.4, 6.5, and 6.6.

6.4.1 StraightTimoBeam Program List

```
% StraightTimoBeam.m
% Solving Straight Timoshenko Beam Problems
% Timoshenko Beam Theory by using NURBS shape functions
clear all; clc;
%ne = 1;                  % Element Division Table 6.4
%ne = 2;                  % Element Division Table 6.5
ne = 4;                   % Element Division Table 6.6
% Material Data
Data(1) = 2e6;            % Young's Modulus E
Data(2) = 2.0/ne;         % Length Le
Data(3) = 0.3; % Poisson's Ratio nu
Data(4) = 5/6; % Shear Correction Coefficient kp
Data(5) = Data(1)/2/(1+Data(3)); % Shear Modulus G = E/2/(1+nu)
% Euler-Bernoulli by using large value ex. G=G*1e5;
Data(6) = 0.2*0.2; % Cross Section Area A
Data(7) = 0.2*0.2^3/12; % Second Moment of Inertia I
Data(8) = 1500; % Density rho
Data(9) = 1; % Order Polynomial Degree of U npa
Data(10)= 3; % Order Polynomial Degree of V npb
Data(11)= 2; % Order Polynomial Degree of Th npc
Data(12)= 4; % Gauss Points ngp
% Jacobi and Gauss data
L = Data(2);
Jac = L/2;
ngp = Data(12);
Gpw = Legendre(ngp);
% Global Matrices
KV = zeros(6+3*(ne-1),6+3*(ne-1));
MV = zeros(6+3*(ne-1),6+3*(ne-1));
FV = zeros(6+3*(ne-1),1);
% Distributed loading qy = 1 N/m
```

```
Lvec = [0 -1.0 0.0]; % note:will be replaced by other loading cases
for ii = 1:ne
    % Constructing Element Matrices
    [KMC,MMC,FVC] = KMTimoStraight(Data,Jac,Lvec,Gpw);
    % Assembling matrices
    if ii==1 % only the first element
      KV(1:6,1:6)=KMC(:,:);
      MV(1:6,1:6)=MMC(:,:);
      FV(1:6,1) =FVC(:,1);
    else
      n1=1+3*(ii-1);
      n2=6+3*(ii-1);
      KV(n1:n2,n1:n2)=KV(n1:n2,n1:n2)+KMC(:,:);
      MV(n1:n2,n1:n2)=MV(n1:n2,n1:n2)+MMC(:,:);
      FV(n1:n2,1) =FV(n1:n2,1)+FVC(:,1);
    end
end % end of ne element sub division ii
% Eliminating fixed DOFs for cantilever support Table 6.4
% KV=KV([4:end],:); % extracting free DOF rows
% KV=KV(:,[4:end]); % extracting free DOF columns
% MV=MV([4:end],:); % extracting free DOF rows
% MV=MV(:,[4:end]); % extracting free DOF columns
% FV=FV([4:end],:); % extracting free DOF rows
% Loading at free end of the beam
%Fload = [0; -1; 0]; % Py = 1 N Case 1
%Fload = [0; 0; 1]; % Py = 1 N Case 2
%Fload=FV; % Distributed loading qy = 1 N/m Case 3
% Table 6.4 ******************************************
% Eliminating fixed DOFs for simply supported beam Table 6.5
% KV=KV([3:end-3,end],:); % extracting free DOF rows
% KV=KV(:,[3:end-3,end]); % extracting free DOF columns
% MV=MV([3:end-3,end],:); % extracting free DOF rows
% MV=MV(:,[3:end-3,end]); % extracting free DOF columns
% FV=FV([3:end-3,end],:); % extracting free DOF rows
% Loading at mid-span of the beam
%FV = [0; 0; -1; 0; 0]; % Py = 1 N Case 1
%FV = [0; 0; 0; 1; 0]; % Mz = 1 N Case 2
%Fload=FV; % Distributed loading qy = 1 N/m Case 3
% Table 6.5 ******************************************
% Static Analysis for Table 6.4 and 6.5 *******************
% disp('Loading Vector - Static Analysis');
% disp(Fload);
% % Static Analysis
% disp('Displacement Vector - Static Analysis');
% Uvec=KV\Fload;
```

```
% disp(Uvec(:)*1000);
% Static Analysis for Table 6.4 and 6.5 ******************
% Free Vibration Analysis Table 6.3 ***************************
% Eliminating fixed DOFs for simply supported Table 6.6
% KV=KV([3,5,6,9],:); % extracting free DOF rows 2 elem
% KV=KV(:,[3,5,6,9]); % extracting free DOF columns
% MV=MV([3,5,6,9],:); % extracting free DOF rows
% MV=MV(:,[3,5,6,9]); % extracting free DOF columns 2 elem
% KV=KV([3,5,6,8,9,12],:); % extracting free DOF rows 3 elem
% KV=KV(:,[3,5,6,8,9,12]); % extracting free DOF columns
% MV=MV([3,5,6,8,9,12],:); % extracting free DOF rows
% MV=MV(:,[3,5,6,8,9,12]); % extracting free DOF columns 3 elem
KV=KV([3,5,6,8,9,11,12,15],:); % extracting free DOF rows 4 elem
KV=KV(:,[3,5,6,8,9,11,12,15]); % extracting free DOF columns
MV=MV([3,5,6,8,9,11,12,15],:); % extracting free DOF rows
MV=MV(:,[3,5,6,8,9,11,12,15]); % extracting free DOF columns 4 elem
% Free Vibration Analysis (Eigenvalue Analysis)
[vec, lam2] = eig(KV,MV);
% Normalized Natural Frequency
lambda=sort(diag(sqrt(lam2)));
ni = size(KV,1);
if ni>6
  ni = 6;
end
omegaN=zeros(ni,1);
omegaTI=[5.1163; 19.5514; 41.1912; 97.6713; 162.7110; 196.6081];
for i=1:ni
    omegaN(i,1) = lambda(i)/omegaTI(i,1);
end
disp('First Six Normalized Natural Frequencies');
disp(omegaN(1:ni,1));
disp(omegaTI(1:ni,1));
% Free Vibration Analysis Table 6.6 ***************************
```

6.4.2 KMTimoStraight Function List

```
function [KMC,MMC,FVC] = KMTimoStraight(Data,Jac,Lvec,Gpw);
% KMTimoStraight.m
% Constructing Stiffness and Mass Matrices of a Straight Beam Element
% Followed by condensation and swapping to the standard DOFs
% Based on the Timoshenko Beam Theory by using NURBS shape functions
E = Data(1);                % Young's Modulus E
```

```
L = Data(2);              % Length Le
nu = Data(3);             % Poisson's Ratio nu
kp = Data(4);             % Shear Correction Coefficient kp
G = Data(5);              % Shear Modulus G = E/2/(1+nu)
% Euler-Bernoulli by using large value ex. G=G*1e5;
A = Data(6);              % Cross Section Area A
I = Data(7);              % Second Moment of Inertia I
rho = Data(8);            % Density rho
npa = Data(9);            % Order Polynomial Degree of U npa
npb = Data(10);           % Order Polynomial Degree of V npb
npc = Data(11);           % Order Polynomial Degree of Th npc
ngp = Data(12);           % Gauss Points ngp
dim = npa+1+npb+1+npc+1;
Kmtx = zeros(dim,dim);
Mmtx = zeros(dim,dim);
Fvec = zeros(dim,1);
knota = [-ones(1,npa+1) ones(1,npa+1)]; knota = [knota ones(1,npa)];
nba = size(knota,2)-npa-1; nna = nba-npa;
wta = ones(nba-npa);
% NURBS for v displacement data
knotb = [-ones(1,npb+1) ones(1,npb+1)]; knotb = [knotb ones(1,npb)];
nbb = size(knotb,2)-npb-1; nnb = nbb-npb;
wtb = ones(nbb-npb);
% NURBS for th displacement data
knotc = [-ones(1,npc+1) ones(1,npc+1)]; knotc = [knotc ones(1,npc)];
nbc = size(knotc,2)-npc-1; nnc = nbc-npc;
wtc = ones(nbc-npc);
Su0 = zeros(nba-npa,ngp); Su1 = zeros(nba-npa,ngp);
Sv0 = zeros(nbb-npb,ngp); Sv1 = zeros(nbb-npb,ngp);
St0 = zeros(nbb-npb,ngp); St1 = zeros(nbb-npb,ngp);
for i=1:ngp
 t = Gpw(i,1); % U displacement
 DNurbsbasis = DNurbsLeibnitz(npa,knota,wta,t);
 Su0(1:nba-npa,i)=DNurbsbasis(1:nba-npa,npa+5); % non-derivative
 Su1(1:nba-npa,i)=DNurbsbasis(1:nba-npa,npa+7); % 1st-derivative
end
for i=1:ngp
 t = Gpw(i,1); % V displacement
 DNurbsbasis = DNurbsLeibnitz(npb,knotb,wtb,t);
 Sv0(1:nbb-npb,i)=DNurbsbasis(1:nbb-npb,npb+5); % non-derivative
 Sv1(1:nbb-npb,i)=DNurbsbasis(1:nbb-npb,npb+7); % 1st-derivative
end
for i=1:ngp
   t = Gpw(i,1); % Theta rotation
   DNurbsbasis = DNurbsLeibnitz(npc,knotc,wtc,t);
   St0(1:nbc-npc,i)=DNurbsbasis(1:nbc-npc,npc+5); % non-derivative
```

```
  St1(1:nbc-npc,i)=DNurbsbasis(1:nbc-npc,npc+7); % 1st-derivative
end
% K11
for i=1:ngp
  for j=1:nba-npa
    for k=1:nba-npa
        Kmtx(j,k) = Kmtx(j,k)+E*A*Su1(j,i)/Jac*Su1(k,i)/Jac*Jac*Gpw
        (i,2);
    end
  end
end
% K22
for i=1:ngp
  for j=1:nnb
    for k=1:nnb
        Kmtx(j+nna,k+nna) = Kmtx(j+nna,k+nna)...
            +kp*G*A*Sv1(j,i)/Jac*Sv1(k,i)/Jac*Jac*Gpw(i,2);
    end
  end
end
% K23
for i=1:ngp
    for j=1:nnb
      for k=1:nnc
          Kmtx(j+nna,k+nna+nnb) = Kmtx(j+nna,k+nna+nnb)...
              -kp*G*A*Sv1(j,i)/Jac*St0(k,i)*Jac*Gpw(i,2);
      end
    end
end
% K32
for i=1:ngp
  for j=1:nnc
      for k=1:nnb
          Kmtx(j+nna+nnb,k+nna) = Kmtx(j+nna+nnb,k+nna)...
              -kp*G*A*St0(j,i)*Sv1(k,i)/Jac*Jac*Gpw(i,2);
      end
  end
end
% K33
for i=1:ngp
  for j=1:nnc
    for k=1:nnc
        Kmtx(j+nna+nnb,k+nna+nnb) = Kmtx(j+nna+nnb,k+nna+nnb)...
            +kp*G*A*St0(j,i)*St0(k,i)*Jac*Gpw(i,2)...
            +E*I*St1(j,i)/Jac*St1(k,i)/Jac*Jac*Gpw(i,2);
```

```
    end
  end
end
% Mass matrix M
% M11
for i=1:ngp
  for j=1:nna
    for k=1:nna
       Mmtx(j,k) = Mmtx(j,k)+rho*A*Su0(j,i)*Su0(k,i)*Jac*Gpw(i,2);
    end
  end
end
% M22
for i=1:ngp
  for j=1:nnb
    for k=1:nnb
       Mmtx(j+nna,k+nna) = Mmtx(j+nna,k+nna)...
           +rho*A*Sv0(j,i)*Sv0(k,i)*Jac*Gpw(i,2);
    end
  end
end
% M33
for i=1:ngp
  for j=1:nnc
      for k=1:nnc
          Mmtx(j+nna+nnb,k+nna+nnb) = Mmtx(j+nna+nnb,
k+nna+nnb)...
             +rho*I*St0(j,i)*St0(k,i)*Jac*Gpw(i,2);
      end
  end
end
% Loading Vector F
% Pu
for i=1:ngp
  for j=1:nna
      Fvec(j,1) = Fvec(j,1)+Lvec(1)*Su0(j,i)*Jac*Gpw(i,2);
  end
end
% Pv
for i=1:ngp
  for j=1:nnb
      Fvec(nna+j,1) = Fvec(nna+j,1)+Lvec(2)*Sv0(j,i)*Jac*Gpw(i,2);
  end
end
% Moment
```

```
for i=1:ngp
  for j=1:nnc
      Fvec(nna+nnb+j,1) = Fvec(nna+nnb+j,1)+Lvec(3)*St0(j,i)
      *Jac*Gpw(i,2);
  end
end
% Condensation
if npa ~= 1
 i=2; j=nna-1; k=nna+nnb+nnc;
 [KMC,MMC,FVC] = Condensation(Kmtx,Mmtx,Fvec,i,j,k); % Condensing u
else
 KMC = Kmtx; MMC = Mmtx; FVC = Fvec;
end
if npb ~= 1
 i=4; j=2+nnb-1; k=2+nnb+nnc;
 [KMC,MMC,FVC] = Condensation(KMC,MMC,FVC,i,j,k); % Condensing v
else
 KMC = Kmtx; MMC = Mmtx; FVC = Lvec;
end
if npc ~= 1
 i=6; j=2+2+nnc-1; k=2+2+nnc;
 [KMC,MMC,FVC] = Condensation(KMC,MMC,FVC,i,j,k); % Condensing theta
else
 KMC = Kmtx; MMC = Mmtx; FVC = Lvec;
end
% Swapping the matrices
KMC=KMC([1,3,5,2,4,6],:); % row
KMC=KMC(:,[1,3,5,2,4,6]); % column
MMC=MMC([1,3,5,2,4,6],:); % row
MMC=MMC(:,[1,3,5,2,4,6]); % column
FVC=FVC([1,3,5,2,4,6],:); % row
return
```

Reference

Cazzani A, Stochino F, Turco E (2016) On the whole spectrum of Timoshenko beams. Part I a theoretical revisitation. Z Angew Math Phys 67(24):1–30

Further Reading

Lee U (2009) Spectral element method in structural dynamics. Wiley, Singapore

Chapter 7
Circular Curved Beam Element Examples

7.1 Static Problem

Consider circular curved Timoshenko beams with two kinds of supports, clamped-clamped and pinned-pinned, shown in Fig. 7.1. The beam has geometry and property depicted in the figure.

The beams are subjected to several loading conditions. Two elements with three Gauss integration points are used in the static analysis. The degrees of polynomial used in the NURBS functions are $p_u = 3$, $p_v = 3$, and $p_\theta = 3$.

The general displacements computed are normalized as follows:

$$\bar{u}_C = \frac{u_C}{\alpha R}; \quad \bar{v}_C = \frac{v_C}{\alpha R}; \quad \bar{\theta}_C = \frac{\theta_C}{\omega} \tag{7.1}$$

The results of static analyses are compared with the results from Litewka and Rakowski (1998). We can see from Table 7.1 that the results are in close agreement with the reference. In the reference, four-element divisions are used in the analyses.

7.2 Free Vibration Problem

Consider the circular curved beam with different supports as shown in Fig. 7.2 for a free vibration problem. We need to divide the beam element into sub-elements to get several mode shapes and frequencies (ω) of the vibration.

Four Gauss integration points are necessary to integrate the mass matrix. The degrees of polynomial used in the NURBS functions are $p_u = 3$, $p_v = 3$, and $p_\theta = 3$.

The results of computed natural frequencies (mode i) are non-dimensionalized by using the following equation:

B.S. Gan, *An Isogeometric Approach to Beam Structures*,
DOI 10.1007/978-3-319-56493-7_7

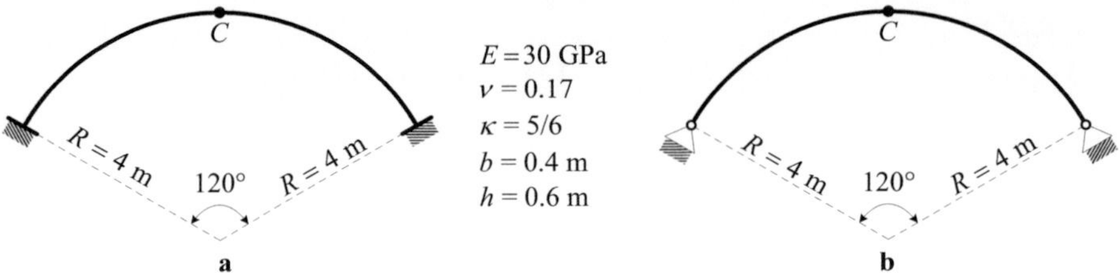

Fig. 7.1 Circular curved beam problem: (**a**) clamped-clamped supports and (**b**) pinned-pinned supports

Table 7.1 Results of circular curved Timoshenko beam in static analysis problems

No.	Circular curved Beam	Normalized	Litewka and Rakowski (1998)	Present
1	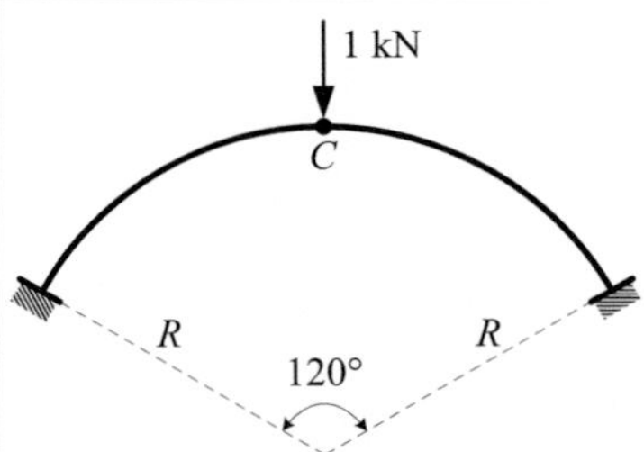	$\bar{u}_C\,[\times 10^{-6}]$	0.0000	0.0000
		$\bar{v}_C\,[\times 10^{-6}]$	−0.2488	−0.2472
		$\bar{\theta}_C\,[\times 10^{-6}]$	0.0000	0.0000
2	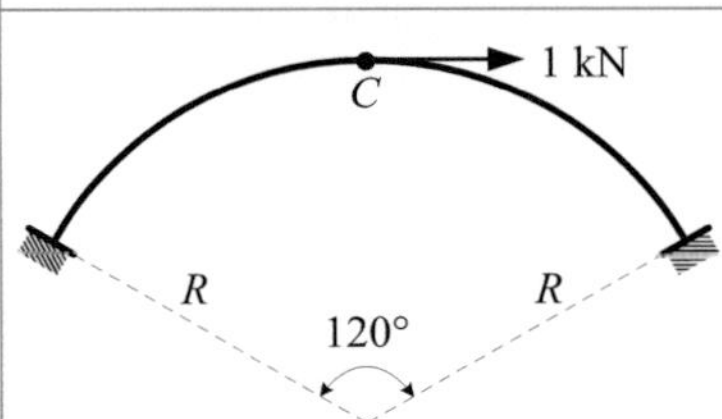	$\bar{u}_C\,[\times 10^{-6}]$	0.1252	0.1246
		$\bar{v}_C\,[\times 10^{-6}]$	0.0000	0.0000
		$\bar{\theta}_C\,[\times 10^{-6}]$	−0.3796	−0.3798
3	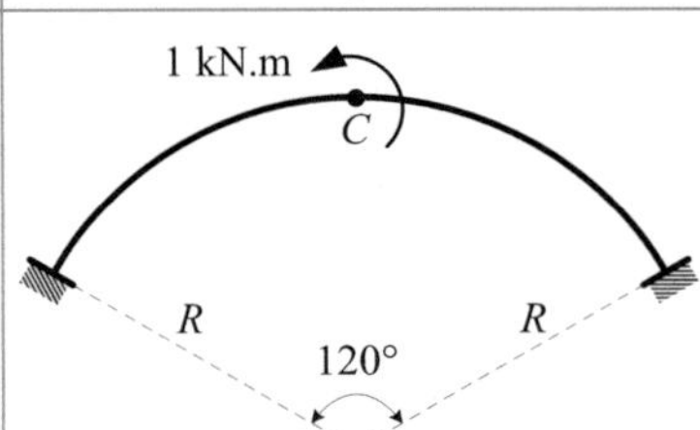	$\bar{u}_C\,[\times 10^{-6}]$	−0.0949	−0.0949
		$\bar{v}_C\,[\times 10^{-6}]$	0.0000	0.0000
		$\bar{\theta}_C\,[\times 10^{-6}]$	1.0824	1.0818
4	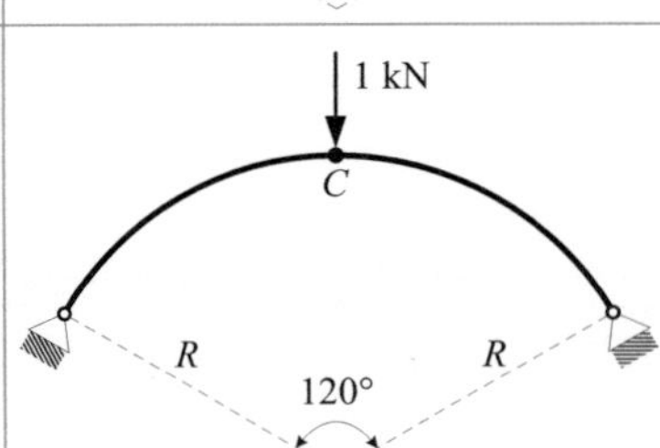	$\bar{u}_C\,[\times 10^{-6}]$	0.0000	0.0000
		$\bar{v}_C\,[\times 10^{-6}]$	−0.3047	−0.2797
		$\bar{\theta}_C\,[\times 10^{-6}]$	0.0000	0.0000

(continued)

Table 7.1 (continued)

No.	Circular curved Beam	Normalized	Litewka and Rakowski (1998)	Present
5	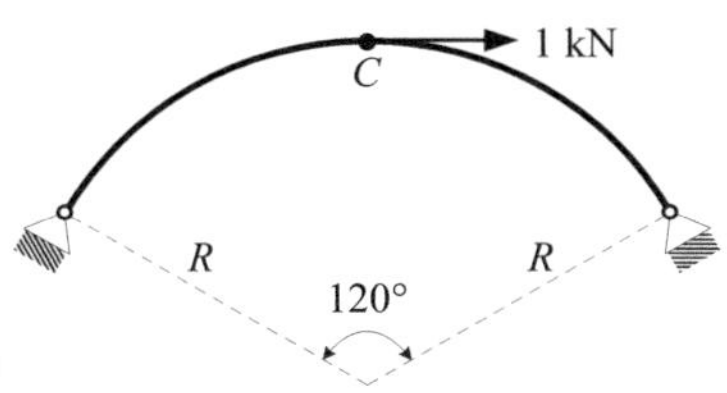	$\bar{u}_C\left[\times10^{-6}\right]$	0.2884	0.2876
		$\bar{v}_C\left[\times10^{-6}\right]$	0.0000	0.0000
		$\bar{\theta}_C\left[\times10^{-6}\right]$	0.8064	0.8066
6	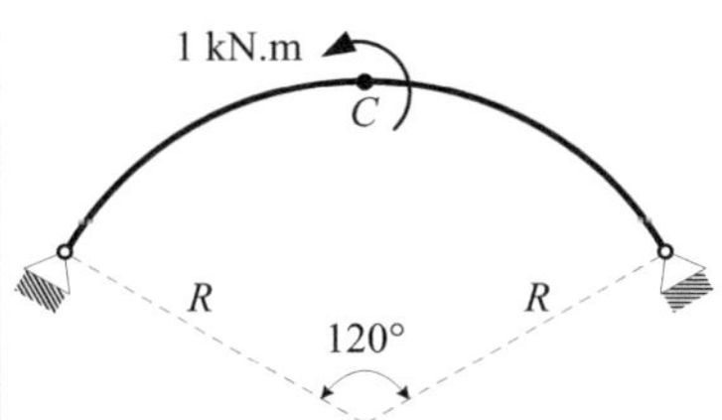	$\bar{u}_C\left[\times10^{-6}\right]$	0.2016	0.2016
		$\bar{v}_C\left[\times10^{-6}\right]$	0.0000	0.0000
		$\bar{\theta}_C\left[\times10^{-6}\right]$	1.3613	1.3611

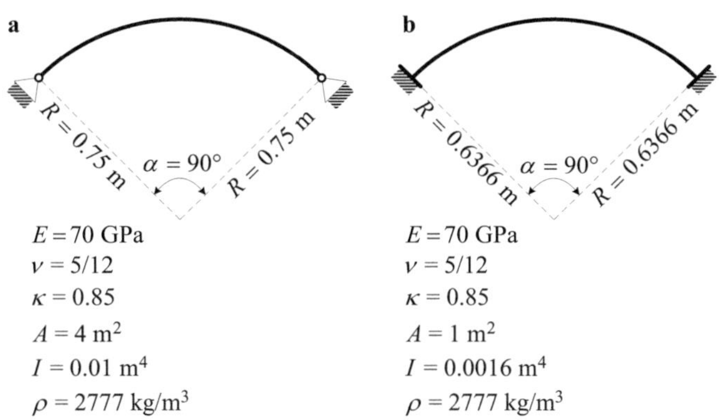

Fig. 7.2 Circular curved beam problem: (**a**) pinned-pinned supports and (**b**) clamped-clamped supports

$$\lambda_i = \omega_i(\alpha R)^2\sqrt{\frac{\rho A}{EI}} \tag{7.2}$$

In the first example as shown in Fig. 7.2a, a pinned-pinned circular curved beam is considered. The results of the first ten non-dimensionalized λ_i are listed in Table 7.2. We can observe a very close agreement between the present approach and those published in the references.

In the second example as shown in Fig. 7.2b, a clamped-clamped circular curved beam is investigated. The results of the first ten non-dimensionalized λ_i are listed in Table 7.3. Similarly, we can observe a very close agreement between the present approach and those reported in the references.

Table 7.2 Non-dimensional frequencies of a pinned-pinned circular curved beam (Fig. 7.2a)

	λ_i			
Mode i	Eisenberger and Efraim (2001)	Veletsos and Austin (1973)	Yang et al. (2008)	Present
1	29.280	29.61	29.306	29.280
2	33.305	33.01	33.243	33.306
3	67.124	67.24	67.123	67.131
4	79.971	79.60	79.950	79.977
5	107.851	107.7	107.844	107.880
6	143.618	144.5	143.679	143.670
7	156.666	155.2	156.629	156.720
8	190.477	191.3	190.596	190.635
9	225.361	223.7	225.349	225.460
10	234.524	235.3	234.809	234.825

Table 7.3 Non-dimensional frequencies of a clamped-clamped circular curved beam (Fig. 7.2b)

	λ_i			
Mode i	Eisenberger and Efraim (2001)	Veletsos and Austin (1973)	Yang et al. (2008)	Present
1	36.703	36.81	36.657	36.703
2	42.264	42.44	42.289	42.264
3	82.233	82.50	82.228	82.243
4	84.491	84.30	84.471	84.495
5	122.305	122.5	122.298	122.341
6	154.945	155.1	154.998	154.990
7	168.203	167.7	168.174	168.271
8	204.472	–	204.599	204.642
9	238.992	–	238.973	239.088
10	249.011	249.6	249.320	249.326

7.3 Program Lists for Solving Circular Curved Timoshenko Beams

MATLAB code CircularTimoBeam.m and functions Condensation.m (Sect. 5.5.2), Legendre.m (Sect. 2.3.2), KMTimoCircular.m (Sect. 7.3.2), and DNurbsLeibnitz.m (Sect. 1.3.14) create element stiffness and mass matrices of a circular curved Timoshenko beam numerically by using NURBS functions. The stiffness and mass matrices are reduced to a two-node six-degree-of-freedom beam element and solved for static and free vibration analyses. The results are presented in Tables 7.1, 7.2, and 7.3.

7.3.1 CircularTimoBeam Program List

```
% CircularTimoBeam.m
% Solving Circular Curved Beam Problems
% Based on the Timoshenko Beam Theory by using NURBS shape
functions
clear all; clc;
%ne = 2; % Element Division ****** Table 7.1
%R = 4; % Radius R ****** Table 7.1
%alpL = 5*pi/6;    % Left Sector of Circle     ****** Table 7.1
%%alpR = pi/6;     % Right Sector of Circle    ****** Table 7.1
ne = 100;          % Element Division *Table 7.2 and Table 7.3
%R = 0.75;         % Radius R                  ****** Table 7.2
R = 0.6366;        % Radius R                  ****** Table 7.3
alpL = 3*pi/4;     % Left Sector of Circle *Table 7.2 and Table 7.3
alpR = pi/4;       % Right Sector of Circle *Table 7.2 and Table 7.3
alp = zeros(ne,2);
% Material Data
%Data(1:6)=[30e9 0.24 0.0072 5/6 0.17 1000];      % ****** Table 7.1
%Data(1:6)=[70e9 4 0.01 0.85 5/12 2777];          % ****** Table 7.2
Data(1:6)=[70e9 1 0.0016 0.85 5/12 2777];         % ****** Table 7.3
Data(7) = Data(1)/2/(1+Data(5)); % Shear Modulus G = E/2/(1+nu)
%Data(7) = Data(7)*1e5;
% Euler-Bernoulli by using large value ex. G=G*1e5;
Data(8) = 3;       % Order Polynomial Degree of U npa
Data(9) = 3;       % Order Polynomial Degree of V npb
Data(10) = 3;      % Order Polynomial Degree of Th npc
Data(11) = 4;      % Gauss Points ngp, *)Table 7.1 use ngp=3
% Global Matrices
KV = zeros(6+3*(ne-1),6+3*(ne-1));
MV = zeros(6+3*(ne-1),6+3*(ne-1));
FV = zeros(6+3*(ne-1),1);
for ii = 1:ne
    alp(ii,1) = alpL-(ii-1)*(alpL-alpR)/ne;
    alp(ii,2) = alpL-(ii)*(alpL-alpR)/ne;
    % Jacobi and Gauss
    Jac = abs(alp(ii,2)-alp(ii,1))/2*R; % Jacobi
    DJac = 1/Jac; % Inversed of Jacobi
    Curv = 1/R; % Constant curvature for a circular curve
    % Constructing Element Matrices
    [KMC,MMC,FVC] = KMTimoCircular(Data,Jac,DJac,Curv);
    % Assembling matrices
    if ii==1 % only the first element
        KV(1:6,1:6)=KMC(:,:);
```

```
        MV(1:6,1:6)=MMC(:,:);
        FV(1:6,1)=FVC(:,:);
    else
        n1=1+3*(ii-1);
        n2=6+3*(ii-1);
        KV(n1:n2,n1:n2)=KV(n1:n2,n1:n2)+KMC(:,:);
        MV(n1:n2,n1:n2)=MV(n1:n2,n1:n2)+MMC(:,:);
        FV(n1:n2,1)=FV(n1:n2,1)+FVC(:,1);
    end
end % end of ne element sub division ii
Normd=1e6/8/pi*3;
Normt=1e6/2/pi*3;
% Table 7.1, No.1-3 ***********************************************
% Eliminating fixed DOFs for fixed-fixed supports
% KV=KV(4:6+3*(ne-1)-3,:); % extracting rows
% KV=KV(:,4:6+3*(ne-1)-3); % extracting columns
% MV=MV(4:6+3*(ne-1)-3,:); % extracting rows
% MV=MV(:,4:6+3*(ne-1)-3); % extracting columns
% FV=zeros((ne-1)*3,1); % No. 1-3
%FV(ne/2*3-1,1)=-1000; % No.1, Py loading
%FV(ne/2*3-2,1)=+1000; % No.2, Px loading
%FV(ne/2*3,1)=+1000; % No.3, M loading
% UV=KV\FV;
% disp('Static Displacements');
% disp([UV((ne/2-1)*3+1)*Normd; UV((ne/2-1)*3+2)*Normd; ...
        % UV((ne/2-1)*3+3)*Normt]);
% Table 7.1, No.4-6 ***********************************************
% Eliminating fixed DOFs for pinned-pinned supports
% KV=KV([3:6+3*(ne-1)-3,end],:); % extracting rows
% KV=KV(:,[3:6+3*(ne-1)-3,end]); % extracting columns
% MV=MV([3:6+3*(ne-1)-3,end],:); % extracting rows
% MV=MV(:,[3:6+3*(ne-1)-3,end]); % extracting columns
% FV=zeros(5,1); % No. 4-6
% FV(3,1)=-1000; % No.4, Py loading
%FV(2,1)=+1000; % No.5, Px loading
%FV(4,1)=+1000; % No.6, M loading
% UV=KV\FV;
% disp('Static Displacements');
% disp([UV(1)*Normt; UV(2)*Normd; UV(3)*Normd; ...
 UV(4)*Normt; UV(5)*Normt]);
% Table 7.2 *******************************************************
% Eliminating fixed DOFs for pinned-pinned support
% KV=KV([3:end-3,end],:); % extracting free DOF rows
% KV=KV(:,[3:end-3,end]); % extracting free DOF columns
% MV=MV([3:end-3,end],:); % extracting free DOF rows
```

```
% MV=MV(:,[3:end-3,end]); % extracting free DOF columns
% Table 7.3 ***************************************************
% Eliminating fixed DOFs for clamped-clamped support
KV=KV(4:end-3,:); % extracting free DOF rows
KV=KV(:,4:end-3); % extracting free DOF columns
MV=MV(4:end-3,:); % extracting free DOF rows
MV=MV(:,4:end-3); % extracting free DOF columns
% Table 7.2 and 7.3 *********************************************
% Free Vibration Analysis (Eigenvalue Analysis)
[vec, lam2] = eig(KV,MV);
% Normalized Natural Frequency
lambda=sort(diag(sqrt(lam2)));
ni = size(KV,1);
if ni>10
  ni = 10;
end
omegaEB=zeros(ni,1);
for i=1:ni
    omegaEB(i,1)=lambda(i)*...
      (alpL-alpR)^2*R^2*sqrt(Data(2)*Data(6)/Data(1)/Data(3));
end
disp('First Ten Normalized Natural Frequencies');
disp(omegaEB(1:ni,1));
% Table 7.2 and 7.3 *********************************************
```

7.3.2 *KMTimoCircular Function List*

```
function [KMC,MMC,FVC] = KMTimoCircular(Data,Jac,DJac,Curv)
% KMTimoCircular.m
% Constructing Stiffness and Mass Matrices of a Circular Beam Element
% Followed by condensation and swapping to the standard DOFs
% Based on the Timoshenko Beam Theory by using NURBS shape functions
E = Data(1);            % Young's Modulus E
A = Data(2);            % Cross Section Area A
I = Data(3);            % Second Moment of Inertia I
kp = Data(4);           % Shear Correction Coefficient kp
nu = Data(5);           % Poisson's Ratio nu
rho = Data(6);          % Density rho
G = Data(7);            % Shear Modulus G
npa = Data(8);          % Order Polynomial Degree of U npa
npb = Data(9);          % Order Polynomial Degree of V npb
npc = Data(10);         % Order Polynomial Degree of Th npc
ngp = Data(11);         % Gauss Points ngp
```

```
dim = npa+1+npb+1+npc+1;
Kmtx = zeros(dim,dim);
Mmtx = zeros(dim,dim);
Lvec = zeros(dim,1);
Gpw = Legendre(ngp);
% NURBS for u displacement data
knota = [-ones(1,npa+1) ones(1,npa+1)]; knota = [knota ones(1,npa)];
nba = size(knota,2)-npa-1; nna = nba-npa;
wta = ones(nba-npa);
% NURBS for v displacement data
knotb = [-ones(1,npb+1) ones(1,npb+1)]; knotb = [knotb ones(1,npb)];
nbb = size(knotb,2)-npb-1; nnb = nbb-npb; wtb = ones(nbb-npb);
% NURBS for th displacement data
knotc = [-ones(1,npc+1) ones(1,npc+1)]; knotc = [knotc ones(1,npc)];
nbc = size(knotc,2)-npc-1; nnc = nbc-npc;
wtc = ones(nbc-npc);
% Gauss data
ngp = Data(11); Gpw = Legendre(ngp);
Su0 = zeros(nba-npa,ngp); Sv0 = zeros(nbb-npb,ngp);
St0 = zeros(nbc-npc,ngp); Su1 = zeros(nba-npa,ngp);
Sv1 = zeros(nbb-npb,ngp); St1 = zeros(nbc-npc,ngp);
for i=1:ngp
  t = Gpw(i,1); % coordinate
  DNurbsbasis = DNurbsLeibnitz(npa,knota,wta,t);
  Su0(1:nba-npa,i)=DNurbsbasis(1:nba-npa,npa+5); % non-derivative
  Su1(1:nba-npa,i)=DNurbsbasis(1:nba-npa,npa+7); % 1st-derivative
end
for i=1:ngp
  t = Gpw(i,1); % coordinate
  DNurbsbasis = DNurbsLeibnitz(npb,knotb,wtb,t);
  Sv0(1:nbb-npb,i)=DNurbsbasis(1:nbb-npb,npb+5); % non-derivative
  Sv1(1:nbb-npb,i)=DNurbsbasis(1:nbb-npb,npb+7); % 1st-derivative
end
for i=1:ngp
  t = Gpw(i,1); % coordinate
  DNurbsbasis = DNurbsLeibnitz(npc,knotc,wtc,t);
  St0(1:nbc-npc,i)=DNurbsbasis(1:nbc-npc,npc+5); % non-derivative
  St1(1:nbc-npc,i)=DNurbsbasis(1:nbc-npc,npc+7); % 1st-derivative
end
% K11
for i=1:ngp
  for j=1:nna
    for k=1:nna
      Kmtx(j,k) = Kmtx(j,k)+E*A*Su1(j,i)*DJac...
              *Su1(k,i)*DJac*Jac*Gpw(i,2)...
             +kp*G*A*Su0(j,i)*Curv*Su0(k,i)*Curv*Jac*Gpw(i,2);
```

```
    end
  end
end
% K12
for i=1:ngp
   for j=1:nna
     for k=1:nnb
         Kmtx(j,k+nna) = Kmtx(j,k+nna)...
             +E*A*Su1(j,i)*DJac*Sv0(k,i)*Curv*Jac*Gpw(i,2)...
             -kp*G*A*Su0(j,i)*Curv*Sv1(k,i)*DJac*Jac*Gpw(i,2);
         Kmtx(k+nna,j) = Kmtx(j,k+nna); % Symmetric K21
     end
   end
end
% K13
for i=1:ngp
   for j=1:nna
     for k=1:nnc
         Kmtx(j,k+nna+nnb) = Kmtx(j,k+nna+nnb)...
             +kp*G*A*Su0(j,i)*Curv*St0(k,i)*Jac*Gpw(i,2);
         Kmtx(k+nna+nnb,j) = Kmtx(j,k+nna+nnb); % Symmetric K31
     end
   end
end
% K22
for i=1:ngp
   for j=1:nnb
     for k=1:nnb
         Kmtx(j+nna,k+nna) = Kmtx(j+nna,k+nna)...
             +E*A*Sv0(j,i)*Curv*Sv0(k,i)*Curv*Jac*Gpw(i,2)...
             +kp*G*A*Sv1(j,i)*DJac*Sv1(k,i)*DJac*Jac*Gpw(i,2);
     end
   end
end
% K23
for i=1:ngp
  for j=1:nnb
     for k=1:nnc
         Kmtx(j+nna,k+nna+nnb) = Kmtx(j+nna,k+nna+nnb)...
             -kp*G*A*Sv1(j,i)*DJac*St0(k,i)*Jac*Gpw(i,2);
         Kmtx(k+nna+nnb,j+nna) = Kmtx(j+nna,k+nna+nnb); % Symmetric K32
     end
   end
end
```

```
% K33
for i=1:ngp
   for j=1:nnc
     for k=1:nnc
         Kmtx(j+nna+nnb,k+nna+nnb) = Kmtx(j+nna+nnb,k+nna+nnb)...
              +kp*G*A*St0(j,i)*St0(k,i)*Jac*Gpw(i,2)...
              +E*I*St1(j,i)*DJac*St1(k,i)*DJac*Jac*Gpw(i,2);
     end
   end
end
% Mass matrix K
% M11
for i=1:ngp
   for j=1:nna
     for k=1:nna
         Mmtx(j,k) = Mmtx(j,k)+rho*A*Su0(j,i)*Su0(k,i)*Jac*Gpw(i,2);
   end
   end
end
% M22
for i=1:ngp
   for j=1:nnb
     for k=1:nnb
         Mmtx(j+nna,k+nna) = Mmtx(j+nna,k+nna)...
              +rho*A*Sv0(j,i)*Sv0(k,i)*Jac*Gpw(i,2);
     end
   end
end
% M33
for i=1:ngp
   for j=1:nnc
     for k=1:nnc
         Mmtx(j+nna+nnb,k+nna+nnb) = Mmtx(j+nna+nnb,k+nna+nnb)...
              +rho*I*St0(j,i)*St0(k,i)*Jac*Gpw(i,2);
     end
   end
end
% Condensation
if npa ~= 1
   i=2; j=nna-1; k=nna+nnb+nnc;
   [KMC,MMC,FVC] = Condensation(Kmtx,Mmtx,Lvec,i,j,k); % Condensing u
end
if npb ~= 1
   i=4; j=2+nnb-1; k=2+nnb+nnc;
   [KMC,MMC,FVC] = Condensation(KMC,MMC,FVC,i,j,k); % Condensing v
end
```

```
if npc ~= 1
  i=6; j=2+2+nnc-1; k=2+2+nnc;
  [KMC,MMC,FVC] = Condensation(KMC,MMC,FVC,i,j,k); % Condensing theta
end
% Swapping the matrices
KMC=KMC([1,3,5,2,4,6],:); % row
KMC=KMC(:,[1,3,5,2,4,6]); % column
MMC=MMC([1,3,5,2,4,6],:); % row
MMC=MMC(:,[1,3,5,2,4,6]); % column
FVC=FVC([1,3,5,2,4,6],:); % row
return
```

References

Eisenberger M, Efraim E (2001) In-plane vibrations of shear deformable curved beams. Int J Num Meth Engrg 52:1221–1234

Litewka P, Rakowski J (1998) The exact thick arch finite element. Comp Struct 68:369–379

Veletsos W, Austin W (1973) Free vibration of arches flexible in shear. J Engrg Mech ASCE 99:735–753

Yang F, Sedaghati R, Esmailzadeh E (2008) Free in-plane vibration of general curved beams using finite element method. J Sound Vib 318:850–867

Chapter 8
General Curved Beam Element Examples

8.1 Parabolic Curved Beam

Consider a parabolic curved beam with pinned-pinned supports as shown in Fig. 8.1 for free vibration problem. The non-dimensional equation for the parabolic curved is defined as

$$\begin{aligned} &y = h(1+\xi)(1-\xi), \qquad -1 \leq \xi \leq +1 \\ &\text{where } \xi = 2x/L - 1 \end{aligned} \tag{8.1}$$

We need to divide the beam element into sub-elements to get several mode shapes and frequencies (ω) of the vibration. Four Gauss integration points are necessary to integrate the mass matrix. The degrees of polynomial used in the NURBS functions are $p_u = 3$, $p_v = 3$, and $p_\theta = 3$. The beam is divided into 20 elements. The arch rise to the span length ratio is defined by $f = h/L$. The slenderness ratio is defined as

$$S_R = L\sqrt{\frac{A}{I}} \tag{8.2}$$

The results of computed natural frequencies (mode i) are non-dimensionalized by using the following equation:

$$\lambda_i = \omega_i L^2 \sqrt{\frac{\rho A}{EI}} \tag{8.3}$$

For the parabolic curved beam free vibration analysis, $f = 0.3$, $S_R = 75$, and $\kappa G/E = 0.3$ are defined. The results of the first ten non-dimensionalized λ_i are listed in Table 8.1. We can observe very close agreement between the present approach and those published in the references.

B.S. Gan, *An Isogeometric Approach to Beam Structures*,
DOI 10.1007/978-3-319-56493-7_8

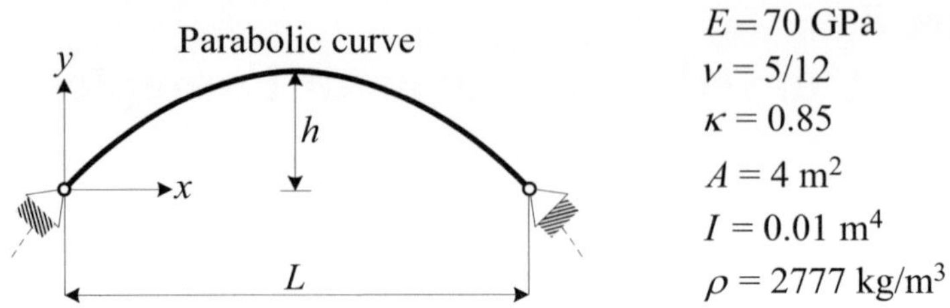

Fig. 8.1 Free vibration problem of a parabolic curved beam with pinned-pinned supports

Table 8.1 Non-dimensional frequencies of a pinned-pinned parabolic curved beam

	λ_i			
Mode i	Veletsos and Austin (1973)	Yang et al. (2008)	Luu et al. (2015)	Present
1	21.83	21.759	21.759	21.758
2	56.00	55.493	55.493	55.501
3	102.3	100.701	100.701	100.788
4	113.4	113.302	113.302	113.414
5	–	–	157.168	157.539
6	–	–	208.969	209.360
7	–	–	224.470	225.469
8	–	–	294.916	297.429
9	–	–	375.630	380.817
10	–	–	411.114	413.366

8.2 Elliptic Curved Beam

Consider an elliptic curved beam with pinned-clamped supports as shown in Fig. 8.2 for free vibration problem. The non-dimensional equation for the elliptic curved is defined as

$$y = b_2\sqrt{1 - \left\{1 - \frac{[\xi + 1 - b_1 Cos\,(a)]}{4b_1}\right\}^2} - b_2 Sin\,(a), \qquad -1 \le \xi \le +1$$

where

$$\xi = (2x/L - 1), \;\; b_1 = \varepsilon + 0.5, \;\; a = \pi - Cos^{-1}(0.5/b_1), \;\; b_2 = h/[1 - Sin\,(a)] \tag{8.4}$$

We need to divide the beam element into sub-elements to get several mode shapes and frequencies (ω) of the vibration. Four Gauss integration points are necessary to integrate the mass matrix. The degrees of polynomial used in the NURBS functions are $p_u = 3$, $p_v = 3$, and $p_\theta = 3$. The beam is divided into 20 elements. The arch rise to the span length ratio is defined by $f = h/L$. The slenderness ratio is defined in (8.2).

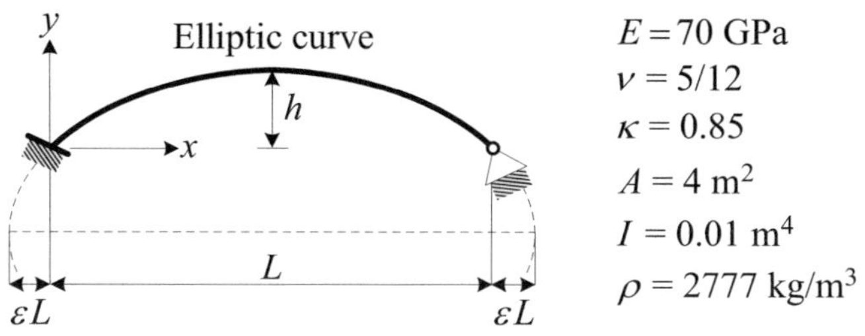

Fig. 8.2 Free vibration problem of an elliptic curved beam with pinned-clamped supports

Table 8.2 Non-dimensional frequencies of a clamped-pinned elliptic curved beam

Mode	λ_i			
i	Veletsos and Austin (1973)	Yang et al. (2008)	Luu et al. (2015)	Present
1	35.25	34.892	34.900	34.896
2	57.11	56.766	56.778	56.780
3	83.00	81.420	81.513	81.498
4	128.2	124.288	124.898	124.565
5	–	–	155.629	155.720
6	–	–	189.888	187.349
7	–	–	264.924	254.530
8	–	–	291.999	292.712
9	–	–	364.547	328.191
10	–	–	432.310	404.479

The results of computed natural frequencies (mode i) are non-dimensionalized by using the Eq. (8.3).

For the elliptic curved beam free vibration analysis, $f = 0.2$, $S_R = 50$, $\varepsilon = 0.5$, and $\kappa G/E = 0.3$ are defined. The results of the first ten non-dimensionalized λ_i are listed in Table 8.2. We can observe very close agreement between the present approach and those published in the references.

8.3 Sinusoidal Curved Beam

Consider a sinusoidal curved beam with clamped-clamped supports as shown in Fig. 8.3 for free vibration problem. The non-dimensional equation for the sinusoidal curved is defined as

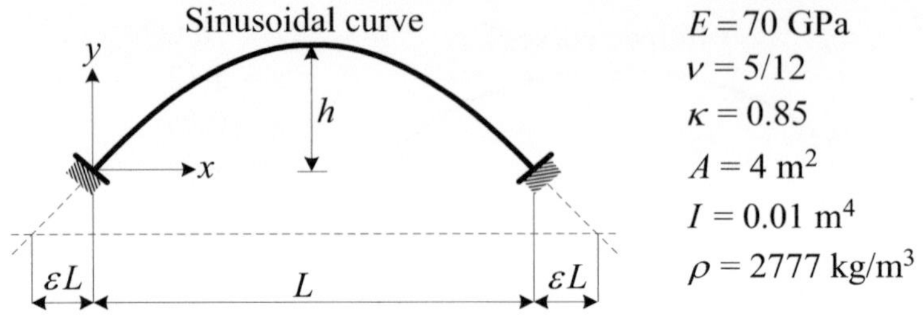

Fig. 8.3 Free vibration problem of a sinusoidal curved beam with clamped-clamped supports

Table 8.3 Non-dimensional frequencies of a clamped-clamped sinusoidal curved beam

	λ_i		
Mode *i*	Veletsos and Austin (1973)	Yang et al. (2008)	Present
1	56.30	56.083	56.088
2	66.14	66.047	66.068
3	114.3	113.406	113.464
4	181.7	179.264	179.490
5	–	–	264.874
6	–	–	311.424
7	–	–	363.879
8	–	–	473.102
9	–	–	595.667
10	–	–	619.024

$$y = h - c_1 + c_1 Sin\left(c_2\left[\frac{\xi+1}{4}\right] + \varepsilon c_2\right), \qquad -1 \le \xi \le +1$$

where

$$\xi = (2x/L - 1), \quad c_2 = \frac{\pi}{1+2\varepsilon}, \quad c_1 = \frac{h}{[1 - Sin\,(\varepsilon c_2)]} \tag{8.5}$$

We need to divide the beam element into sub-elements to get several mode shapes and frequencies (ω) of the vibration. Four Gauss integration points are necessary to integrate the mass matrix. The degrees of polynomial used in the NURBS functions are $p_u = 3$, $p_v = 3$, and $p_\theta = 3$. The beam is divided into 20 elements. The arch rise to the span length ratio is defined by $f = h/L$. The slenderness ratio is defined in (8.2).

The results of computed natural frequencies (mode *i*) are non-dimensionalized by using Eq. (8.3).

For the elliptic curved beam free vibration analysis, $f = 0.1$, $S_R = 100$, $\varepsilon = 0.5$, and $\kappa G/E = 0.3$ are defined. The results of the first ten non-dimensionalized λ_i are listed in Table 8.3. We can observe very close agreement between the present approach and those published in the references.

8.4 Program Lists for Solving General Curved Timoshenko Beams

MATLAB code GeneralTimoBeam.m and functions Condensation.m (Sect. 5.5.2), Legendre.m (Sect. 2.3.2), KMTimoGeneral.m (Sect. 8.4.2), and DNurbsLeibnitz.m (Sect. 1.3.14) create element stiffness and mass matrices of general curved Timoshenko beams, parabolic, elliptic, and sinusoidal curved beams, numerically by using NURBS functions. The stiffness and mass matrices are reduced to a two-node six degree-of-freedom beam element and solved for free vibration analyses. The results are compared with other works and presented in Tables 8.1, 8.2, and 8.3.

8.4.1 GeneralTimoBeam Program List

```
% GeneralTimoBeam.m - 20 Elements
% Solving General Curved Beam Free Vibration Problems
% Timoshenko Beam Theory by using NURBS shape functions
clear all; clc;
% Element Division - Span and Length Ratio - Slenderness Ratio SR
%ne = 20; f = 0.3; SR = 75; % *********************** Table 8.1
%ne = 20; f = 0.2; SR = 50; % *********************** Table 8.2
ne = 20; f = 0.1; SR =100; % *********************** Table 8.3
% Material Data
Data(1) = 70e9;     % Young's Modulus E
Data(2) = 4;        % Cross Section Area A
Data(3) = 0.01;     % Second Moment of Inertia I
Data(4) = 0.85;     % Shear Correction Coefficient kp
Data(5) = 5/12;     % Poisson's Ratio nu
Data(6) = 2777;     % Density rho
Data(7) = Data(1)/2/(1+Data(5)); % Shear Modulus G = E/2/(1+nu)
Data(8) = 3;        % Order Polynomial Degree of U npa
Data(9) = 3;        % Order Polynomial Degree of V npb
Data(10)= 3;        % Order Polynomial Degree of Th npc
Data(11)= 4;        % Gauss Points ngp
npa = Data(8);
% Gauss data
ngp = Data(11); Gpw = Legendre(ngp);
% Global Matrices
KV = zeros(6+3*(ne-1),6+3*(ne-1));
MV = zeros(6+3*(ne-1),6+3*(ne-1));
L = SR*sqrt(Data(3)/Data(2));
Le = L/ne;
xi=zeros(1,npa+1); yi=zeros(1,npa+1);
```

```
% Knot vector
knot=[-1*ones(1,npa+1)  1*ones(1,npa+1)]; % Uniform Rational
Bspline
knot=[knot max(knot)*ones(1,npa)];
nb=size(knot,2)-npa-1;          % max number of basis functions
% Parameters
wt=[ones(1,npa+1)];
tmi=zeros(ngp,npa+1);
% Jacobi and Curvature
Jac=zeros(ngp,1);
DJac=zeros(ngp,1);
Curv=zeros(ngp,1);
% Table 8.2
%epsi=0.5; b1=epsi+0.5; a=pi-acos(0.5/b1); b2=f*L/(1-sin(a));
% Table 8.3
epsi=0.5; c2=pi/(1+2*epsi); c1=f*L/(1-sin(epsi*c2));
for ii = 1:ne
    xi0 = (ii-1)*Le;
    for ip=1:npa+1
      xi(1,ip)=xi0+(ip-1)*Le/npa;
      ksi=2*xi(1,ip)/L-1;
      %yi(1,ip)=f*L*(1+ksi)*(1-ksi);%*******************Table 8.1
     %yi(1,ip)=b2*sqrt(1-(1-((ksi+1)/2-b1*cos(a))/b1)^2)-b2*sin(a);8.2
     yi(1,ip)=f*L-c1+c1*sin(c2*(ksi+1)/2+epsi*c2);%******Table 8.3
    end
    xyi=transpose([xi yi]);
    t=[-1:2/npa:1];
    nt=size(t,2);
    for i=0:nb-npa-1
      for dt=1:nt
        Nurbsbasis = Nurbs(npa,knot,wt,t(dt));
        tmi(dt,i+1)=Nurbsbasis(i+1,npa+2);
      end
    end
    % [tm] matrix
    tm=[tmi zeros(nb-npa,nb-npa);zeros(nb-npa,nb-npa) tmi];
    % Control Points Px's and Py's calculation
    pxy=tm\xyi;
    px=pxy(1:nt); % n+1 X control points
    py=pxy(nt+1:2*nt); % n+1 Y control points
    % Jacobi and Curvature calculation
    for it=1:nt
      DNurbsbasis = DNurbsLeibnitz(npa,knot,wt,t(it));
      dx=0; dy=0; ddy=0; ddx=0;
```

```
      for ib=1:nb-npa                          % Number of basis function
         dx=dx+DNurbsbasis(ib,npa+7)*px(ib);          % dx
         dy=dy+DNurbsbasis(ib,npa+7)*py(ib);          % dy
         ddx=ddx+DNurbsbasis(ib,npa+8)*px(ib);        % dx
         ddy=ddy+DNurbsbasis(ib,npa+8)*py(ib);        % dy
      end
      Jac(it) =sqrt(dx^2+dy^2);
      DJac(it) =1/Jac(it);
      Curv(it) =abs(dx*ddy-dy*ddx)/Jac(it)^3;
    end
    % Constructing Element Matrices
    [KMC,MMC,FVC] = KMTimoGeneral(Data,Jac,DJac,Curv,Gpw);
    % Assembling matrices
    if ii==1 % only the first element
      KV(1:6,1:6)=KMC(:,:);
      MV(1:6,1:6)=MMC(:,:);
    else
       n1=1+3*(ii-1);
       n2=6+3*(ii-1);
       KV(n1:n2,n1:n2)=KV(n1:n2,n1:n2)+KMC(:,:);
       MV(n1:n2,n1:n2)=MV(n1:n2,n1:n2)+MMC(:,:);
    end
end % end of ne element sub division ii
% Eliminating fixed DOFs for pinned-pinned support ******** Ta-
ble 8.1
% KV=KV([3:end-3,end],:); % extracting free DOF rows
% KV=KV(:,[3:end-3,end]); % extracting free DOF columns
% MV=MV([3:end-3,end],:); % extracting free DOF rows
% MV=MV(:,[3:end-3,end]); % extracting free DOF columns *** Ta-
ble 8.1
% Eliminating fixed DOFs for clamped-pinned support ******* Ta-
ble 8.2
% KV=KV([4:end-3,end],:); % extracting free DOF rows
% KV=KV(:,[4:end-3,end]); % extracting free DOF columns
% MV=MV([4:end-3,end],:); % extracting free DOF rows
% MV=MV(:,[4:end-3,end]); % extracting free DOF columns *** Ta-
ble 8.2
% Eliminating fixed DOFs for clamped-clamped support ****** Ta-
ble 8.3
KV=KV([4:end-3],:); % extracting free DOF rows
KV=KV(:,[4:end-3]); % extracting free DOF columns
MV=MV([4:end-3],:); % extracting free DOF rows
MV=MV(:,[4:end-3]); % extracting free DOF columns ********* Ta-
ble 8.3
% Free Vibration Analysis (Eigenvalue Analysis)
```

```
[vec, lam2] = eig(KV,MV);
% Normalized Natural Frequency
lambda=sort(diag(sqrt(lam2)));
ni = size(KV,1);
if ni>10
   ni = 10;
end
omegaEB=zeros(ni,1);
for i=1:ni
  omegaEB(i,1)=lambda(i)*...
      L^2*sqrt(Data(2)*Data(6)/Data(1)/Data(3));
end
disp('First Ten Normalized Natural Frequencies');
disp(omegaEB(1:ni,1));
```

8.4.2 KMTimoGeneral Function List

```
function [KMC,MMC,FVC] = KMTimoGeneral(Data,Jac,DJac,Curv,Gpw)
% KMTimoGeneral.m
% Constructing Stiffness and Mass Matrices of a General Beam Element
% Followed by condensation and swapping to the standard DOFs
% Based on the Timoshenko Beam Theory by using NURBS shape functions
E = Data(1);                % Young's Modulus E
A = Data(2);                % Cross Section Area A
I = Data(3);                % Second Moment of Inertia I
kp = Data(4);               % Shear Correction Coefficient kp
nu = Data(5);               % Poisson's Ratio nu
rho = Data(6);              % Density rho
G = Data(7);                % Shear Modulus G
npa = Data(8);              % Order Polynomial Degree of U npa
npb = Data(9);              % Order Polynomial Degree of V npb
npc = Data(10);             % Order Polynomial Degree of Th npc
ngp = Data(11);             % Gauss Points ngp
% NURBS for u displacement data
knota = [-ones(1,npa+1) ones(1,npa+1)]; knota = [knota ones(1,npa)];
nba    = size(knota,2)-npa-1; nna = nba-npa; wta = ones(nba-npa);
% NURBS for v displacement data
knotb = [-ones(1,npb+1) ones(1,npb+1)]; knotb = [knotb ones(1,npb)];
nbb    = size(knotb,2)-npb-1; nnb = nbb-npb; wtb = ones(nbb-npb);
% NURBS for th displacement data
knotc = [-ones(1,npc+1) ones(1,npc+1)]; knotc = [knotc ones(1,npc)];
nbc    = size(knotc,2)-npc-1; nnc = nbc-npc; wtc = ones(nbc-npc);
```

```
dim    = npa+1+npb+1+npc+1;
Lvec   = zeros(dim,1);
Kmtx   = zeros(dim,dim);
Mmtx   = zeros(dim,dim);
% Shape functions
Su0     = zeros(nba-npa,ngp); Sv0 = zeros(nbb-npb,ngp);
St0     = zeros(nbc-npc,ngp); Su1 = zeros(nba-npa,ngp);
Sv1     = zeros(nbb-npb,ngp); St1 = zeros(nbc-npc,ngp);
for i=1:ngp
     t = Gpw(i,1); % coordinate
     DNurbsbasis = DNurbsLeibnitz(npa,knota,wta,t);
     Su0(1:nba-npa,i)=DNurbsbasis(1:nba-npa,npa+5); % non-derivative
     Su1(1:nba-npa,i)=DNurbsbasis(1:nba-npa,npa+7); % 1st-derivative
end
for i=1:ngp
    t = Gpw(i,1); % coordinate
    DNurbsbasis = DNurbsLeibnitz(npb,knotb,wtb,t);
    Sv0(1:nbb-npb,i)=DNurbsbasis(1:nbb-npb,npb+5); % non-derivative
    Sv1(1:nbb-npb,i)=DNurbsbasis(1:nbb-npb,npb+7); % 1st-derivative
end
for i=1:ngp
     t = Gpw(i,1); % coordinate
     DNurbsbasis = DNurbsLeibnitz(npc,knotc,wtc,t);
     St0(1:nbc-npc,i)=DNurbsbasis(1:nbc-npc,npc+5); % non-derivative
     St1(1:nbc-npc,i)=DNurbsbasis(1:nbc-npc,npc+7); % 1st-derivative
end
% K11
for i=1:ngp
    for j=1:nna
        for k=1:nna
            Kmtx(j,k)=Kmtx(j,k)+E*A*Su1(j,i)*DJac(i)*Su1(k,i)*DJac(i)...
                 *Jac(i)*Gpw(i,2)+kp*G*A*Su0(j,i)*Curv(i)*Su0(k,i)...
                *Curv(i)*Jac(i)*Gpw(i,2);
        end
    end
end
% K12
for i=1:ngp
    for j=1:nna
        for k=1:nnb
            Kmtx(j,k+nna) = Kmtx(j,k+nna)...
            +E*A*Su1(j,i)*DJac(i)*Sv0(k,i)*Curv(i)*Jac(i)*Gpw(i,2)...
            -kp*G*A*Su0(j,i)*Curv(i)*Sv1(k,i)*DJac(i)*Jac(i)*Gpw(i,2);
                 Kmtx(k+nna,j) = Kmtx(j,k+nna); % Symmetric K21
            end
```

```
      end
    end
% K13
for i=1:ngp
     for j=1:nna
          for k=1:nnc
                Kmtx(j,k+nna+nnb) = Kmtx(j,k+nna+nnb)...
                  +kp*G*A*Su0(j,i)*Curv(i)*St0(k,i)*Jac(i)*Gpw(i,2);
                Kmtx(k+nna+nnb,j) = Kmtx(j,k+nna+nnb); % Symmetric K31
          end
 end
     end
% K22
for i=1:ngp
   for j=1:nnb
     for k=1:nnb
        Kmtx(j+nna,k+nna) = Kmtx(j+nna,k+nna)...
            +E*A*Sv0(j,i)*Curv(i)*Sv0(k,i)*Curv(i)*Jac(i)*Gpw(i,2)...
           +kp*G*A*Sv1(j,i)*DJac(i)*Sv1(k,i)*DJac(i)*Jac(i)*Gpw(i,2);
     end
   end
end
% K23
for i=1:ngp
   for j=1:nnb
      for k=1:nnc
         Kmtx(j+nna,k+nna+nnb) = Kmtx(j+nna,k+nna+nnb)...
               -kp*G*A*Sv1(j,i)*DJac(i)*St0(k,i)*Jac(i)*Gpw(i,2);
         Kmtx(k+nna+nnb,j+nna) = Kmtx(j+nna,k+nna+nnb); % Symmetric K32
      end
   end
end
% K33
for i=1:ngp
   for j=1:nnc
      for k=1:nnc
         Kmtx(j+nna+nnb,k+nna+nnb) = Kmtx(j+nna+nnb,k+nna+nnb)...
         +kp*G*A*St0(j,i)*St0(k,i)*Jac(i)*Gpw(i,2)...
         +E*I*St1(j,i)*DJac(i)*St1(k,i)*DJac(i)*Jac(i)*Gpw(i,2);
      end
   end
end
% Mass matrix K
% M11
for i=1:ngp
```

```
  for j=1:nna
    for k=1:nna
      Mmtx(j,k)=Mmtx(j,k)+rho*A*Su0(j,i)*Su0(k,i)*Jac(i)*Gpw(i,2);
    end
  end
end
% M22
for i=1:ngp
  for j=1:nnb
    for k=1:nnb
      Mmtx(j+nna,k+nna) = Mmtx(j+nna,k+nna)...
      +rho*A*Sv0(j,i)*Sv0(k,i)*Jac(i)*Gpw(i,2);
    end
  end
end
% M33
for i=1:ngp
  for j=1:nnc
    for k=1:nnc
      Mmtx(j+nna+nnb,k+nna+nnb) = Mmtx(j+nna+nnb,k+nna+nnb)...
      +rho*I*St0(j,i)*St0(k,i)*Jac(i)*Gpw(i,2);
    end
  end
end
% Condensation
if npa ~= 1
 i=2; j=nna-1; k=nna+nnb+nnc;
 [KMC,MMC,FVC] = Condensation(Kmtx,Mmtx,Lvec,i,j,k); % Condensing u
end
if npb ~= 1
 i=4; j=2+nnb-1; k=2+nnb+nnc;
 [KMC,MMC,FVC] = Condensation(KMC,MMC,FVC,i,j,k); % Condensing v
end
if npc ~= 1
 i=6; j=2+2+nnc-1; k=2+2+nnc;
 [KMC,MMC,FVC] = Condensation(KMC,MMC,FVC,i,j,k); % Condensing theta
end
% Swapping the matrices
KMC=KMC([1,3,5,2,4,6],:); % row
KMC=KMC(:,[1,3,5,2,4,6]); % column
MMC=MMC([1,3,5,2,4,6],:); % row
MMC=MMC(:,[1,3,5,2,4,6]); % column
FVC=FVC([1,3,5,2,4,6],:); % row
return
```

References

Luu AT, Kim NI, Lee J (2015) Isogeometric vibration analysis of free-form Timoshenko curved beams. Meccanica 50(1):169–187

Veletsos W, Austin W (1973) Free vibration of arches flexible in shear. J Engrg Mech ASCE 99:735–753

Yang F, Sedaghati R, Esmailzadeh E (2008) Free in-plane vibration of general curved beams using finite element method. J Sound Vib 318:850–867

Chapter 9
Free Curved Beam Element Examples

9.1 M-Shape Curved Beam

Consider an M-shape curved beam with pinned-pinned supports subjected to vertical loading $P = 25$ kN at the mid-span of the structure (Fig. 9.1). The properties of the beam are shown in the figure. The static and free vibration analyses were conducted.

Eight Gauss integration points are used to integrate the stiffness and mass matrices. The degrees of polynomial used in the NURBS functions are $p_u = 7$, $p_v = 7$, and $p_\theta = 7$. The beam is divided into two elements. The element matrices are condensed to a two-node six degree-of-freedom beam element.

The equation of this curved beam shape is defined by

$$y = x\ \mathrm{Sin}\,(x) + 3.02721 \quad ; \quad -4 \le x \le +4 \tag{9.1}$$

The knot vectors are given as

$$\begin{aligned} \Xi_{ne=1} &= [-1\ -1\ -1\ -1\ -1\ -1\ -1\ -1\ 1\ 1\ 1\ 1\ 1\ 1\ 1\ 1] \\ \Xi_{ne=2} &= [-1\ -1\ -1\ -1\ -1\ -1\ -1\ -1\ 1\ 1\ 1\ 1\ 1\ 1\ 1\ 1] \end{aligned} \tag{9.2}$$

The control points can be calculated from sample points given by using Eq. (1.14). The sample points can be calculated from (9.1), where the center of coordinates is taken at the point of loading. The sample points on the curve, weight, and local coordinate ξ are given in Table 9.1.

The results of computed vertical displacement and the first five natural frequencies are given in Table 9.2.

B.S. Gan, *An Isogeometric Approach to Beam Structures*,
DOI 10.1007/978-3-319-56493-7_9

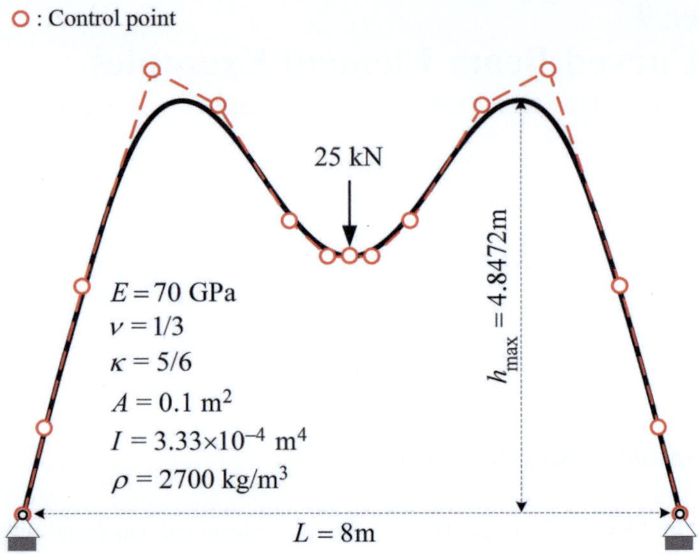

Fig. 9.1 M-shape curved beam with pinned-pinned supports

Table 9.1 Input data for the M-shape curved beam with pinned-pinned supports

x_i	y_i	w_i	ξ_i
−4.00	0.00000	1	−1.00
−3.40	2.15837	1	−0.70
−2.80	3.96518	1	−0.40
−2.00	4.84580	1	0.00
−1.40	4.40684	1	0.30
−0.60	3.36600	1	0.70
−0.20	3.06694	1	0.90
0.00	3.02721	1	1.00
0.00	3.02721	1	−1.00
0.20	3.06694	1	−0.90
0.60	3.36600	1	−0.70
1.40	4.40684	1	−0.30
2.00	4.84580	1	0.00
2.80	3.96518	1	0.40
3.40	2.15837	1	0.70
4.00	0.00000	1	1.00

9.2 Spring-Shape Curved Beam

Consider a spring-shape curved cantilever beam subjected to horizontal loading $P = 15$ kN at the free end of the structure (Fig. 9.2). The properties of the beam are shown in the figure. The static and free vibration analyses were conducted.

Table 9.2 Result of static and free vibration analyses of an M-shape curved beam

Static Analysis	Vertical displacement at the mid-span $v = -0.2649$ mm Rotational displacement at pinned supports $\theta = 0.0000011$ radians				
Free Vibration Mode-i	1	2	3	4	5
ω_i (rad/s)	68.8	243.4	249.0	313.7	1202.5
Eigenvector	−0.1233	−1.0000	−1.0000	−0.3329	0.0439
	−1.0000	0.0000	0.1402	0.0000	−0.0587
	0.0000	−0.1303	0.0000	1.0000	−0.0000
	0.1750	−0.0000	−0.0833	−0.0000	−1.0000
	−0.1233	1.0000	−1.0000	0.3329	0.0439

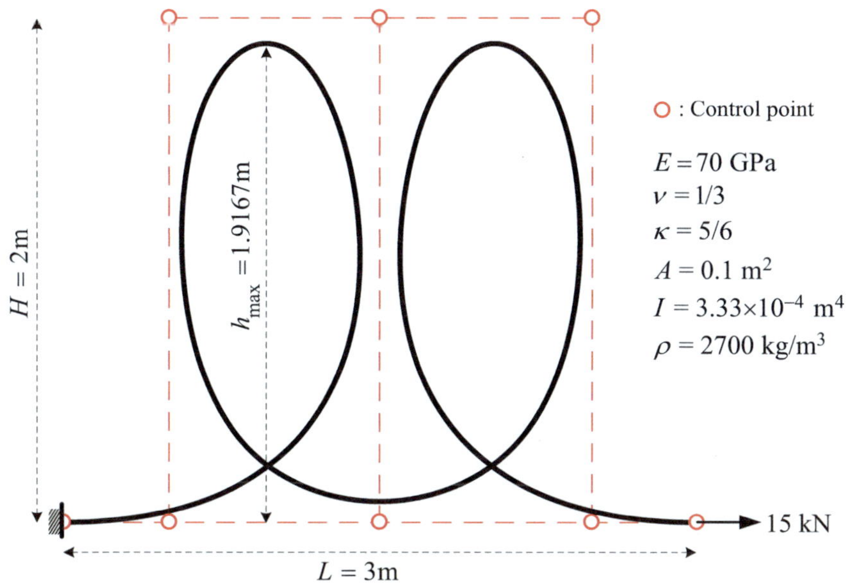

Fig. 9.2 Spring-shape curved cantilever beam

Ten Gauss integration points are used to integrate the stiffness and mass matrices. The degree of polynomial used in the NURBS functions are $p_u = 3$, $p_v = 3$ and $p_\theta = 3$. The beam is modeled by using one element. The element matrices are condensed to a two-node six degree-of-freedom beam element.

The knot vector is given as

$$\boldsymbol{\Xi} = \begin{bmatrix} -1 & -1 & -1 & -1 & -1 & -1 & -1 & -1 & -1 & -1 \\ 1 & 1 & 1 & 1 & 1 & 1 & 1 & 1 & 1 & 1 \end{bmatrix} \tag{9.3}$$

The control points of the curve and weights are given in Table 9.3.

Table 9.3 Input data for the Spring-shape curved cantilever beam

CPx_i	CPy_i	w_i
0.00	0.00	1
1.50	0.00	1
1.50	2.00	1
0.50	2.00	1
0.50	0.00	1
2.50	0.00	1
2.50	2.00	1
1.50	2.00	1
1.50	0.00	1
3.00	0.00	1

Table 9.4 Result of static and free vibration analyses of a Spring-shape curved beam

Static analysis	Horizontal displacement at the free end $u = 1.0554$ mm Vertical displacement at the free end $v = -0.2877$ mm Rotational displacement at the free end $\theta = 0.0011$ radians		
Free Vibration Mode-i	1	2	3
ω_i (rad/s)	234.6	1186.9	5438.2
Eigenvector	0.9774	0.2785	−0.0494
	−0.2422	−0.7568	0.1717
	1.0000	1.0000	−1.0000

The results of computed vertical displacement and the first three natural frequencies are given in Table 9.4.

9.3 Program Lists for Solving Free Curved Timoshenko Beams

MATLAB code FreeTimoBeam.m and functions Condensation.m (Sect. 5.5.2), Legendre.m (Sect. 2.3.2), KMTimoGeneral.m (Sect. 8.4.2), and DNurbsLeibnitz.m (Sect. 1.3.14) create element stiffness and mass matrices of free curved Timoshenko beams, M-shape, spring-shape free curved beams, numerically by using NURBS functions. The stiffness and mass matrices are reduced to a two-node six degree-of-freedom beam element and solved by static analysis and free vibration analyses. The results are presented in Tables 9.2 and 9.4.

9.3.1 *FreeTimoBeam Program List*

```
% FreeTimoBeam.m - Free curves
% Solving Free Curved Beam Free Vibration Problems
% Timoshenko Beam Theory by using NURBS shape functions
clear all; clc;
% Element Division (this program works only for either ne=1 or ne=2)
%ne = 2; % *********************** Table 9.2
ne = 1; % *********************** Table 9.4
% Material Data
Data(1) = 70e9;       % Young's Modulus E
Data(2) = 0.1;        % Cross Section Area A
Data(3) = 10/30000;   % Second Moment of Inertia I
Data(4) = 5/6;        % Shear Correction Coefficient kp
Data(5) = 1/3;        % Poisson's Ratio nu
Data(6) = 2700;       % Density rho
Data(7) = Data(1)/2/(1+Data(5)); % Shear Modulus G = E/2/(1+nu)
% Order Polynomial Degree of U npa, V npb, Th npc, Gauss Points ngp
if ne==2
  Data(8) = 7; Data(9) = 7; Data(10)= 7; Data(11)= 8; %**** Table 9.2
else
  Data(8) = 3; Data(9) = 3; Data(10)= 3; Data(11)= 4; %**** Table 9.4
end
% NURBS for u displacement data
npa = Data(8);
knota = [-ones(1,npa+1) ones(1,npa+1)]; knota = [knota ones(1,npa)];
nba = size(knota,2)-npa-1; nna = nba-npa;
wta = ones(nba-npa);
% Gauss data
ngp = Data(11); Gpw = Legendre(ngp);
% Global Matrices
KV = zeros(6+3*(ne-1),6+3*(ne-1));
MV = zeros(6+3*(ne-1),6+3*(ne-1));
nb = size(knota,2)-npa-1;          % max number of basis functions
% Parameters
tmi=zeros(ngp,nb-npa);
% Jacobi and Curvature
Jac=zeros(ngp,1);
DJac=zeros(ngp,1);
Curv=zeros(ngp,1);
cosphi=zeros(ngp,1);
sinphi=zeros(ngp,1);
```

```
if ne == 1                                 % ****** Table 9.4 ******
   % Control Points and Ksi
   px(1,1:10)=[0 1.5 1.5 0.5 0.5 2.5 2.5 1.5 1.5 3];
   py(1,1:10)=[0 0 2 2 0 0 2 2 0 0];
   wt=ones(1,10);
   % Knot vector
   t=-1:2/9:1;                             % ****** Table 9.4 ******
else                                       % ****** Table 9.2 ******
   % Sampling Points and Ksi
   xi(1,1:8)=[-4 -3.4 -2.8 -2 -1.4 -0.6 -0.2 0];
   xi(2,1:8)=[ 0 0.2 0.6 1.4 2 2.8 3.4 4];
   yi(1,1:8)=[0 2.15837 3.96518 4.84580 4.40684 3.36600 ...
   3.06694 3.02721];
   yi(2,1:8)=[3.02721 3.06694 3.36600 4.40684 4.84580 3.96518 ...
                                                 2.15837 0];
   t(1,1:8)=[-1 -0.7 -0.4 0 0.3 0.7 0.9 1];
   t(2,1:8)=[-1 -0.9 -0.7 -0.3 0 0.4 0.7 1];
   wt=ones(1,nb-npa);                      % ****** Table 9.2 ******
end
nt=size(t,2);
for ii = 1:ne
   if ne == 2                              % ****** Table 9.2 ******
     xyi=transpose([xi(ii,:) yi(ii,:)]);
     for i=1:nb-npa
       for dt=1:nt
         Nurbsbasis = DNurbsLeibnitz(npa,knota,wt,t(ii,dt));
         tmi(dt,i)=Nurbsbasis(i,npa+5);
       end
     end
     % [tm] matrix
     tm=[tmi zeros(nb-npa,nb-npa);zeros(nb-npa,nb-npa) tmi];
     % Control Points Px's and Py's calculation
     pxy=tm\xyi;
     px=pxy(1:nt); % n+1 X control points
     py=pxy(nt+1:2*nt); % n+1 Y control points
   end                                     % ****** Table 9.2 ******
   % Jacobi and Curvature calculation
   for it=1:ngp
     ksi=Gpw(it,1);
     DNurbsbasis = DNurbsLeibnitz(npa,knota,wt,ksi);
     dx=0; dy=0; ddy=0; ddx=0;
     for ib=1:nb-npa % Basis function
     dx=dx+DNurbsbasis(ib,npa+7)*px(ib);   % dx
     dy=dy+DNurbsbasis(ib,npa+7)*py(ib);   % dy
     ddx=ddx+DNurbsbasis(ib,npa+8)*px(ib); % dx
     ddy=ddy+DNurbsbasis(ib,npa+8)*py(ib); % dy
   end
```

```
    Jac(it) =sqrt(dx^2+dy^2);
    DJac(it) =1/Jac(it);
    Curv(it) =abs(dx*ddy-dy*ddx)/Jac(it)^3;
  end
  % Constructing Element Matrices
  [KMC,MMC,FVC] = KMTimoGeneral(Data,Jac,DJac,Curv,Gpw);
  % Coordinate Transformation
  for it=1:nt-1:nt
    DNurbsbasis = DNurbsLeibnitz(npa,knota,wt,t(ii,it));
    dx=0; dy=0;
    for ib=1:nb-npa                          % Basis function
      dx=dx+DNurbsbasis(ib,npa+7)*px(ib); % dx
      dy=dy+DNurbsbasis(ib,npa+7)*py(ib); % dy
  end
  cosphi(it) = dx/sqrt(dx*dx+dy*dy);
  sinphi(it) = dy/sqrt(dx*dx+dy*dy);
  end
  TM = diag(ones(6,1),0);
  TM(1,1) = cosphi(1); TM(2,2) = TM(1,1);
  TM(1,2) = sinphi(1); TM(2,1) =-TM(1,2);
  TM(4,4) = cosphi(nt); TM(5,5) = TM(4,4);
  TM(4,5) = sinphi(nt); TM(5,4) =-TM(4,5);
  KMC = TM'*KMC*TM;
  MMC = TM'*MMC*TM;
  % Assembling matrices
  if ii==1 % only the first element
    KV(1:6,1:6)=KMC(:,:);
    MV(1:6,1:6)=MMC(:,:);
  else
    n1=1+3*(ii-1);
    n2=6+3*(ii-1);
    KV(n1:n2,n1:n2)=KV(n1:n2,n1:n2)+KMC(:,:);
    MV(n1:n2,n1:n2)=MV(n1:n2,n1:n2)+MMC(:,:);
  end
end % end of ne element sub division ii
if ne == 2
   % Eliminating fixed DOFs for pinned-pinned support ****** Table 9.2
   KV=KV([3:end-3,end],:); % extracting free DOF rows
   KV=KV(:,[3:end-3,end]); % extracting free DOF columns
   MV=MV([3:end-3,end],:); % extracting free DOF rows
   MV=MV(:,[3:end-3,end]); % extracting free DOF columns
   FV=[0 0 -25000 0 0]'; %**** Table 9.2
else
   %Eliminating fixed DOFs for clamped-free support ***** Table 9.4
   KV=KV([end-2:end],:); % extracting unconstrained DOF rows
```

```
   KV=KV(:,[end-2:end]); % extracting unconstrained DOF columns
   MV=MV([end-2:end],:); % extracting unconstrained DOF rows
   MV=MV(:,[end-2:end]); % extracting unconstrained DOF columns
   FV=[15000 0 0]'; %***** Table 9.4
end
% Free Vibration Analysis (Eigenvalue Analysis)
Vdis=KV\FV;
disp('Displacements');
disp(Vdis);
% Free Vibration Analysis (Eigenvalue Analysis)
[vec, lam2] = eig(KV,MV);
modal = [diag(sqrt(lam2)/1000)' ; vec]';
% Natural Frequency
disp('Natural Frequencies(X1000)rad/s and Eigenvectors');
transpose(sortrows(modal))
```

Appendix A
Straight Beam Element Matrices

Abstract
In this appendix, element stiffness and mass matrices of a straight beam are provided for validation of the results from numerical integration methods.

Keywords
Stiffness matrix; Mass matrix; Straight beam element; Euler-Bernoulli; Timoshenko

A.1 Stiffness Matrix of Euler-Bernoulli Straight Beam Element

The stiffness element matrix of the Euler-Bernoulli beam is given as

$$\mathbf{K} = \begin{bmatrix} \frac{EA}{L} & 0 & 0 & -\frac{EA}{L} & 0 & 0 \\ 0 & \frac{12EI}{L^3} & \frac{6EI}{L^2} & 0 & -\frac{12EI}{L^3} & \frac{6EI}{L^2} \\ 0 & \frac{6EI}{L^2} & \frac{4EI}{L} & 0 & -\frac{6EI}{L^2} & \frac{2EI}{L} \\ -\frac{EA}{L} & 0 & 0 & \frac{EA}{L} & 0 & 0 \\ 0 & -\frac{12EI}{L^3} & -\frac{6EI}{L^2} & 0 & \frac{12EI}{L^3} & -\frac{6EI}{L^2} \\ 0 & \frac{6EI}{L^2} & \frac{2EI}{L} & 0 & -\frac{6EI}{L^2} & \frac{4EI}{L} \end{bmatrix} \tag{A.1}$$

B.S. Gan, *An Isogeometric Approach to Beam Structures*,
DOI 10.1007/978-3-319-56493-7

A.2 Mass Matrix of Euler-Bernoulli Straight Beam Element

For the completeness, the mass element matrices of the Euler-Bernoulli beam can be separated into two kinds of matrices. The first mass element matrix is associated with translational inertia [$\mathbf{M}_{\rho A}$], and the second mass element matrix is associated with the rotatory inertia [$\mathbf{M}_{\rho I}$]. In the Euler-Bernoulli beam theory, the rotatory inertia matrix is neglected. While in the Rayleigh beam theory, the rotatory inertia matrix is taken into account.

$$\mathbf{M} = \mathbf{M}_{\rho A} + \mathbf{M}_{\rho I} \tag{A.2}$$

where

$$\mathbf{M}_{\rho A} = \frac{\rho A L}{420}\begin{bmatrix} 140 & 0 & 0 & 70 & 0 & 0 \\ 0 & 156 & 22L & 0 & 54 & -13L \\ 0 & 22L & 4L^2 & 0 & 13L & -3L^2 \\ 70 & 0 & 0 & 140 & 0 & 0 \\ 0 & 54 & 13L & 0 & 156 & -22L \\ 0 & -13L & -3L^2 & 0 & -22L & 4L^2 \end{bmatrix}$$

$$\mathbf{M}_{\rho I} = \frac{\rho I}{30L}\begin{bmatrix} 0 & 0 & 0 & 0 & 0 & 0 \\ 0 & 36 & 3L & 0 & -36 & 3L \\ 0 & 3L & 4L^2 & 0 & -3L & -L^2 \\ 0 & 0 & 0 & 0 & 0 & 0 \\ 0 & -36 & -3L & 0 & 36 & -3L \\ 0 & 3L & -L^2 & 0 & -3L & 4L^2 \end{bmatrix}$$

A.3 Stiffness Matrix of Timoshenko Straight Beam Element

The stiffness element matrix of the Timoshenko beam is given as

$$\mathbf{K} = C \times \begin{bmatrix} \frac{A}{I}(1+\phi) & 0 & 0 & -\frac{A}{I}(1+\phi) & 0 & 0 \\ 0 & 12 & 6L & 0 & -12 & 6L \\ 0 & 6L & (4+\phi)L^2 & 0 & -6L & (2-\phi)L^2 \\ -\frac{A}{I}(1+\phi) & 0 & 0 & \frac{A}{I}(1+\phi) & 0 & 0 \\ 0 & -12 & -6L & 0 & 12 & -6L \\ 0 & 6L & (2-\phi)L^2 & 0 & -6L & (4+\phi)L^2 \end{bmatrix} \tag{A.3}$$

where

$$C = \frac{EI}{(1+\phi)L^3}; \qquad \phi = \frac{12}{L^2}\left(\frac{EI}{\kappa GA}\right)$$

A.4 Mass Matrix of Timoshenko Straight Beam Element

For convenience, the mass element matrices of the Timoshenko beam are separated into two kinds of matrices. The first mass element matrix is associated with translational inertia $[\mathbf{M}_{\rho A}]$, and the second mass element matrix is associated with the rotatory inertia $[\mathbf{M}_{\rho I}]$.

$$\mathbf{M} = \mathbf{M}_{\rho A} + \mathbf{M}_{\rho I} \tag{A.4}$$

where

$$\mathbf{M}_{\rho A} = \frac{\rho A L}{210(1+\phi)^2} \times \begin{bmatrix} 70(1+\phi)^2 & 0 & 0 & 35(1+\phi)^2 & 0 & 0 \\ 0 & (70\phi^2+147\phi+78) & (35\phi^2+77\phi+44)\frac{L}{4} & 0 & (35\phi^2+63\phi+27) & -(35\phi^2+63\phi+26)\frac{L}{4} \\ 0 & (35\phi^2+77\phi+44)\frac{L}{4} & (7\phi^2+14\phi+8)\frac{L^2}{4} & 0 & (35\phi^2+63\phi+26)\frac{L}{4} & -(7\phi^2+14\phi+6)\frac{L^2}{4} \\ 35(1+\phi)^2 & 0 & 0 & 70(1+\phi)^2 & 0 & 0 \\ 0 & (35\phi^2+63\phi+27) & (35\phi^2+63\phi+26)\frac{L}{4} & 0 & (70\phi^2+147\phi+78) & -(35\phi^2+77\phi+44)\frac{L}{4} \\ 0 & -(35\phi^2+63\phi+26)\frac{L}{4} & -(7\phi^2+14\phi+6)\frac{L^2}{4} & 0 & -(35\phi^2+77\phi+44)\frac{L}{4} & (7\phi^2+14\phi+8)\frac{L^2}{4} \end{bmatrix}$$

$$\mathbf{M}_{\rho I} = \frac{\rho I}{30(1+\phi)^2 L} \times \begin{bmatrix} 0 & 0 & 0 & 0 & 0 & 0 \\ 0 & 36 & -(15\phi-3)L & 0 & -36 & -(15\phi-3)L \\ 0 & -(15\phi-3)L & (10\phi^2+5\phi+4)L^2 & 0 & (15\phi-3)L & (5\phi^2-5\phi-1)L^2 \\ 0 & 0 & 0 & 0 & 0 & 0 \\ 0 & -36 & (15\phi-3)L & 0 & 36 & (15\phi-3)L \\ 0 & -(15\phi-3)L & (5\phi^2-5\phi-1)L^2 & 0 & (15\phi-3)L & (10\phi^2+5\phi+4)L^2 \end{bmatrix}$$

Index

B.S. Gan, *An Isogeometric Approach to Beam Structures*,
DOI 10.1007/978-3-319-56493-7

MIX
Papier aus verantwortungsvollen Quellen
Paper from responsible sources
FSC® C105338

If you have any concerns about our products,
you can contact us on
ProductSafety@springernature.com

In case Publisher is established outside the EU,
the EU authorized representative is:
Springer Nature Customer Service Center GmbH
Europaplatz 3, 69115 Heidelberg, Germany

Printed by Libri Plureos GmbH
in Hamburg, Germany